Geometry Problem Solving for Middle School

Kevin Wang
Kelly Ren
John Lensmire
Wilson Cheung

Copyright © 2016 Areteem Institute

Published by Areteem Institute
www.areteem.org
All rights reserved.

ISBN: 1-944863-05-2
ISBN-13: 978-1-944863-05-0

First printing, October 2016.

Contents

Introduction 7

1 Counting Through Patterns 15
1.1 Example Questions 18
1.2 Quick Response Questions 25
1.3 Practice 27

2 Practice with Measurements 35
2.1 Example Questions 38
2.2 Quick Response Questions 46
2.3 Practice 48

3 A Dance with Angles 57
3.1 Example Questions 61
3.2 Quick Response Questions 68
3.3 Practice 72

Copyright © ARETEEM INSTITUTE. All rights reserved.

4 Magic in Triangles 81
4.1 Example Questions 84
4.2 Quick Response Questions 92
4.3 Practice 96

5 You Are Special, Right? 103
5.1 Example Questions 108
5.2 Quick Response Questions 115
5.3 Practice 117

6 Angles Are Special Too 125
6.1 Example Questions 128
6.2 Quick Response Questions 136
6.3 Practice 137

7 Discovering Areas 143
7.1 Example Questions 145
7.2 Quick Response Questions 155
7.3 Practice 157

8 Conquering Areas 165
8.1 Example Questions 166
8.2 Quick Response Questions 174
8.3 Practice 176

9 Circle Around 185
9.1 Example Questions 190
9.2 Quick Response Questions 198
9.3 Practice 200

10 Circle Back 207
10.1 Example Questions 208
10.2 Quick Response Questions 217
10.3 Practice 219

11	A Whole New Dimension	227
11.1	Example Questions	234
11.2	Quick Response Questions	242
11.3	Practice	244
	Answer Key to Problems	251
	Index	271

Introduction

Do you know what geometry is? If you are in middle school, you may have some idea that it has to do with shapes and you are not far off in assuming so. Truth be told, we all possess some kind of "geometric intuition" even before we take this class subject in school. "Geometric intuition" refers to the fact that when we look at the world around us, we instantly recognize patterns found in nature. Everything, including the sun, plants and the rest of the world around us, can all be seen in terms of mathematics and the shapes found within these can be understood in the context of geometry. If it were not for mathematics, for example, we may not have known that the intricately curved design of the Nautilus shell actually applies the concept of the Fibonacci sequence to perfection.

Geometry is also an essential part of many people's lives, especially when it comes to their work in a specific career. Architectural design might be one of the first application fields of geometry that comes to mind. Architects employ geometric concepts for the practical design of a structure (such as determining the angle from the structure to the ground that will allow for a certain height and weight distribution) as well as the aesthetic look of a structure. Modern development in STEAM (Science, Technology, Engineering, Arts, and Mathematics) fields further stress the importance of mathematics. Geometry is one of the essential subject areas to lay the foundations for students to eventually excel in trending fields such as robotics design, 3D animation, VR (the fascinating world of virtual reality), GPS (Global Positioning System), aeronautical and aerospace engineering (for flight and space vehicles such as those who have designed and currently run the International Space Station) and for innovative automotive engineering (such as autonomous cars to run the way they are designed). Math therefore provides a way for people to make sense of the world!

Although careers may not be your immediate concern, geometry does provide a foundation in problem solving that will develop over time. Geometry provides a natural way to transition from "geometric intuition" to abstract thinking. In our efforts to assist students, this curriculum book provides an invaluable introduction to the mathematical concepts found in geometry. The content of this book presents geometric concepts in an easily accessible way for students in middle school to understand while providing the kind of academic challenge they need to help them improve their critical thinking skills. In this way, the intrinsic "geometric intuition" that students already possess is molded to allow for more abstract thinking, necessary for understanding geometry.

To accomplish this feat, students will find that the problems contained within these pages are related to situations found in their daily lives. By using real-life scenarios to demonstrate mathematics, students will improve their spatial visualization, helping them easily transition to the kind of problem solving that is needed when completing, for instance, a more formal geometric proof. Geometry is a standard high school mathematics course serving as a gateway to more advanced mathematics later in high school and at the collegiate level. Earlier exposure to the basic ideas and vocabulary used in geometry will allow students to be confident in approaching geometry again when taken as a subject in school.

In this book, we present more in-depth problem solving, covering the application of fundamental concepts in areas, angles, surface areas and volumes and how students can readily apply these concepts in their own lives, highlighted with pictures and 3D shapes to illustrate the problems. The book covers in-depth implementation of Common Core Math Standards for geometry that all middle school students are required to understand before entering high school.

Finally, this book serves as a training tool for students interested in participating in and expanding upon their knowledge of middle school math competitions. Among others, the materials covered in the book have a proven record of effectiveness for those students who want to excel in the American Math Competition (AMC) 8, MathCounts, and Division M of the monthly Zoom International Math League (ZIML). Major competition problems are reviewed and explained, allowing for students to realize the multiple ways in which they can maneuver through math problems with ease. Mathematical equations

Copyright © ARETEEM INSTITUTE. All rights reserved.

are shown to be multi-layered and as such a variety of methods must be employed to solve every problem and come to a complete solution. As a result, students improve their problem solving techniques and benefit from their newfound skills in thinking more critically when approaching competition mathematics.

It is our hope at Areteem Institute that you will find this curriculum book helpful in providing a solid foundation at the middle school level for the success of studying high school geometry and beyond. Students will be able to approach geometry with a better understanding of its concepts and build the confidence to solve problems, whether in a school class setting or in the heat of the moment of a math competition. This book serves as a Student Workbook and has an accompanying Solutions Manual with detailed explanations of the problems included in this Workbook. We hope that you enjoy this book!

Common Core and This Book

Teachers and students that are in 6th, 7th, and 8th grade math can use this book to teach and learn mathematical reasoning and problem solving, focusing on concepts in the Common Core Geometry domain.

For reference, a summary of the middle school geometry domain is provided below.

Standard(s)	Cluster
6.G.1-4	Solve real-world and mathematical problems involving area, surface area, and volume.
7.G.1-3	Draw, construct, and describe geometrical figures and describe the relationships between them.
7.G.4-6	Solve real-life and mathematical problems involving angle measure, area, surface area, and volume.
8.G.1-5	Understand congruence and similarity using physical models, transparencies, or geometry software.
8.G.6-8	Understand and apply the Pythagorean Theorem.
8.G.9	Solve real-world and mathematical problems involving volume of cylinders, cones, and spheres.

The start of each chapter summarizes the specific Common Core standards emphasized in the chapter. While the focus of the book is geometry, the problem solving stressed in the exercises allows students to practice in other domains as well. Exercises with calculations allow students to put the standards taught in the "Number System" (NS) and "Expressions and Equations" (EE) domains to good use. Geometric concepts such as similarity also provide excellent examples for use in the "Ratios and Proportional Relationships" (RP) domain.

For more details about the specific standards, clusters, and domains quoted above, see www.corestandards.org/Math where the full Mathematical Standards are available for download.

Copyright © ARETEEM INSTITUTE. All rights reserved.

Introduction

About Areteem Institute

Areteem Institute is an educational institution that develops and provides in-depth and advanced math and science programs for K-12 (Elementary School, Middle School, and High School) students and teachers. Areteem programs are accredited supplementary programs by the Western Association of Schools and Colleges (WASC). Students may attend the Areteem Institute through these options:

- Live and real-time face-to-face online classes with audio, video, interactive online whiteboard, and text chatting capabilities;
- Self-paced classes by watching the recordings of the live classes;
- Topical short video courses for trending math, science, technology, engineering, English, and social studies topics;
- Summer Intensive Camps on prestigious university campuses and Winter Boot Camps;
- Practice with selected daily problems for free, monthly ZIML competition at `ziml.areteem.org`.

The Areteem courses are designed and developed by educational experts and industry professionals to bring real world applications into STEM education. The programs are ideal for students who wish to build their mathematical strength in order to excel academically and eventually win in Math Competitions (AMC, AIME, USAMO, IMO, ARML, MathCounts, Math Olympiad, ZIML, and other math leagues and tournaments, etc.), Science Fairs (County Science Fairs, State Science Fairs, national programs like Intel Science and Engineering Fair, etc.) and Science Olympiad, or purely want to enrich their academic lives by taking more challenges and developing outstanding analytical, logical thinking and creative problem solving skills.

Since 2004 Areteem Institute has been teaching with methodology that is highly promoted by the new Common Core State Standards: stressing the conceptual level understanding of the math concepts, problem solving techniques, and solving problems with real world applications. With the guidance from experienced and passionate professors, students are motivated to explore concepts deeper by identifying an interesting problem, researching it, analyzing it, and using a critical thinking approach to come up with multiple solutions.

Thousands of math students who have been trained at Areteem achieved top honors and earned top awards in major national and international math competitions, including Gold Medalists in the International Math Olympiad (IMO), top winners and qualifiers at the USA Math Olympiad (USAMO/JMO), and AIME, top winners at the Zoom

International Math League (ZIML), and top winners at the MathCounts National. Many Areteem Alumni have graduated from high school and gone on to enter their dream colleges such as MIT, Cal Tech, Harvard, Stanford, Yale, Princeton, U Penn, Harvey Mudd College, UC Berkeley, UCLA, etc. Those who have graduated from colleges are now playing important roles in their fields of endeavor.

Further information about Areteem Institute, as well as updates and errata of this book, can be found online at `http://www.areteem.org`.

Acknowledgments

This book contains many years of collaborative work by the staff of Areteem Institute. This book could not have existed without their efforts. Huge thanks go to the Areteem staff for their contributions, in particular, Cameron Yanoscik for his help with drafting the overview of the introduction and providing editorial update of the book.

The examples and problems in this book were either created by the Areteem staff or adapted from various sources, including other books and online resources. Especially, some good problems from American Mathematics Competitions (AMC) 8 and MATH-COUNTS are chosen as examples to illustrate concepts or problem-solving techniques. The original resources are credited whenever possible. However, it is not practical to list all such resources. We extend our gratitude to the original authors of all these resources.

1. Counting Through Patterns

This chapter introduces the fundamental shapes in geometry, and focuses on recognizing and counting the fundamental shapes in various figures.

The concepts introduced in this chapter directly correspond to Common Core Math Standards as shown in the following table.

6th Grade	6.G.1, 6.EE.6
7th Grade	7.G.1, 7.G.5
8th Grade	8.EE.7, 8.G.1, 8.G.2

In addition to the standards above, problems and concepts in this section will help strengthen understanding of the following domains.

6th Grade	6.G, 6.NS, 6.EE
7th Grade	7.G, 7.NS, 7.EE
8th Grade	8.EE, 8.G

Copyright © ARETEEM INSTITUTE. All rights reserved.

Points

Points are the fundamental building block of geometry. Points have no length, area, volume, etc. It is often useful to think of a point as a 'dot'.

Lines

A line is a straight object that extends forever in both directions. Given two points, there is a unique line that connects the two points.

Example 1.1 Intersecting and Parallel Lines

Two different lines may or may not intersect. If they intersect, they do so at a point. If they do not intersect, we call the lines parallel.

Example 1.2 Angles

When two lines intersect, as in the diagram below

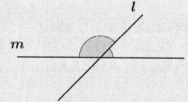

they form angles. Two of the angles are highlighted. We often measure angles in degrees.

Remark

We will study angles in more detail in Chapter 3.

Line Segments and Length

Given two points, the part of the line between the two points is called a line segment. The size of a line segment is called its length.

Shapes and Area

Forms or figures of objects are called shapes. Objects such as triangles, rectangles, circles, etc. are all shapes. The amount of space a shape takes up is called its area.

Perimeter

The length around the outside of a shape is called is perimeter.

Example 1.3 Common Shapes

- Triangles have 3 sides. If all the sides have the same length, we call it an equilateral triangle.
- Parallelograms have 4 sides, with the opposite sides parallel. Rectangles are one special type of parallelogram where all 4 angles are right angles. Another special parallelogram is the rhombus where all 4 sides have the same length.

Congruent

Two shapes that are the same are called congruent. Congruent shapes have the same side lengths and the same angles.

Copyright © ARETEEM INSTITUTE. All rights reserved.

1.1 Example Questions

Example 1.4

How many angles less than 180° are there in the following diagram?

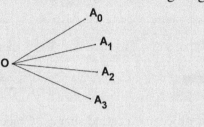

Solution

There are 3 small angles
$$\angle A_0OA_1, \angle A_1OA_2, \angle A_2OA_3,$$
2 medium angles
$$\angle A_0OA_2, \angle A_1OA_3,$$
and 1 large angle
$$\angle A_0OA_3.$$
Therefore, the number of angles in the diagram is
$$3+2+1=6.$$

Example 1.5

(2009 AMC 8) How many non-congruent triangles have vertices at three of the eight points in the array shown below?

. . . .

. . . .

Copyright © ARETEEM INSTITUTE. All rights reserved.

1.1 Example Questions

Solution

We can assume that the base of the triangle is on the bottom four points since a congruent triangle can be formed by reflecting the base of the top four points. There are 3 noncongruent triangles with base length 1, 3 noncongruent triangles with base length 2, and 2 noncongruent triangles with base length 3. Therefore, there is a total of

$$3+3+2=8$$

triangles.

Example 1.6

Count the triangles in each of the following diagrams:

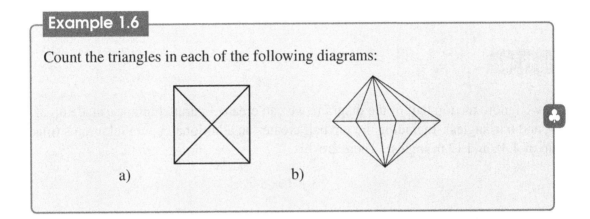

a)   b)

Solution

To count accurately, categorize the triangles.

For example, in a), there are 4 one-piece triangles and 4 two-piece triangles, for a total of 8 triangles.

In b), the top half of the diagram has

$$5+4+3+2+1=15$$

triangles. Similarly the bottom has 15. Lastly there are 5 triangles which overlap the top and bottom for a total of

$$15+15+5=35$$

triangles.

> **Example 1.7**
>
> In the following diagram, $ABCD$ is a parallelogram, and each of the segments in the diagram is parallel to one of $\overline{AB}$, $\overline{AD}$, or $\overline{BE}$. Count the number of parallelograms in the diagram that contain the shaded triangle.
>
>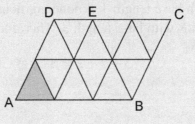

> **Solution**

If we ignore the top half of the diagram, we can create 3 parallelograms (made up of 2, 4, and 6 triangles). Including the top half creates an additional 3 parallelograms (made up of 4, 8, and 12 triangles). There are

$$3+3=6$$

parallelograms containing the shaded region in the diagram.

> **Example 1.8**
>
> (2011 AMC 8) How many rectangles are in this figure?
>
>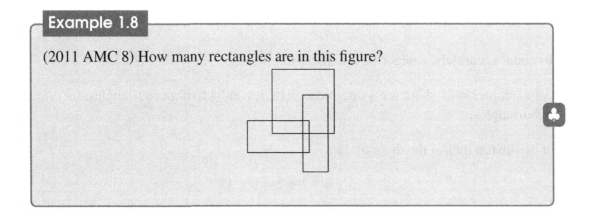

1.1 Example Questions

Solution

The figure can be interpreted in terms of sections. There are 3 rectangles with one section, 5 rectangles with two sections, and 3 with four sections. There are a total of

$$3+5+3 = 11$$

rectangles in the figure.

Example 1.9

Arrange several equilateral triangles, all of whose side lengths are 2cm, to form a long parallelogram, as shown in the diagram. Assume the perimeter of the long parallelogram is 144cm, how many triangles are there?

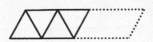

Solution

The left and right sides each has length 2 cm, so the remaining sides of the parallelogram have length 140 cm. As each triangle has a side length of 2 cm, there are

$$140 \div 2 = 70$$

triangles.

Example 1.10

Given a circular disk, use 3 lines to divide the disk into small regions. At most how many regions can there be? What if there are 4 lines?

Solution

When we draw one line in the circle, we create 2 total regions. In drawing further lines, we attempt to have the cuts intersect all of the previous cuts to maximize the number of pieces.

Therefore, for two lines, we have a total of 4 regions. As depicted below, we have 7 total regions for three lines and 11 total regions for four lines.

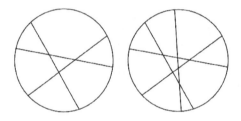

Example 1.11

In the diagram, each side is perpendicular to its adjacent sides, and all small sides have equal length. Given that the perimeter of this diagram is 108cm, find the area of the shape.

Solution

Note that the diagram has the same perimeter as a square with the same dimensions. The square thus has a side length of
$$108 \div 4 = 27$$
cm. As a side of the square is made up of 3 small sides from the diagram, each small side has length
$$27 \div 3 = 9$$

Copyright © ARETEEM INSTITUTE. All rights reserved.

cm. As the diagram is made up of five squares with this side length, the total area of the shape is
$$5 \times 9^2 = 405 \text{cm}^2.$$

Example 1.12

(2004 AMC 8) Thirteen black and six white hexagonal tiles were used to create the figure below. If a new figure is created by attaching a border of white tiles with the same size and shape as the others, what will be the difference between the total number of white tiles and the total number of black tiles in the new figure?

Solution

Note that the first ring around the middle tile has 6 tiles, and the second has 12. From this pattern, the third ring has 18 tiles. Of these tiles,
$$6 + 18 = 24$$
are white and
$$1 + 12 = 13$$
are black. Therefore, the difference is
$$24 - 13 = 11.$$

Example 1.13

Use 4 congruent rectangles to form one big square, as shown. The big square has area 100 cm². Suppose the width of each rectangle is 1cm. What is the perimeter of each rectangle?

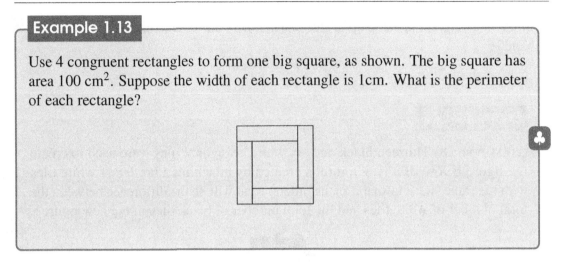

Solution

The side length of the square is 10 cm, so if the width of each rectangle is 1 cm, the length must be 9 cm. Hence the perimeter is

$$2 \times 1 + 2 \times 9 = 20 \text{cm}.$$

1.2 Quick Response Questions

Problem 1.1 The diagram below is used for the following four questions.

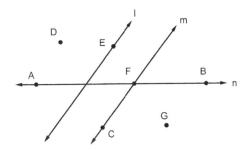

How many points are there in the picture? How many lines?

Problem 1.2 What points are on line n? Are there any points that lie on two lines? Name a point not on any line.

Problem 1.3 Which line is determined by the points C and F? That is, what is another name for the line $\overleftrightarrow{CF}$?

Problem 1.4 What is another way to describe or determine line n?

Problem 1.5 How many triangles are in each of the diagrams below?

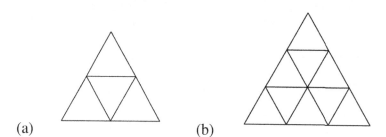

(a) (b)

1.3 Practice

Problem 1.6 In the following diagram, $\angle 1 = \angle 2 = \angle 3$. The sum of the measures of all possible angles in $\angle AOB$ is $180°$. What is the measure of $\angle AOB$?

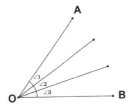

Problem 1.7 In the following diagram, each small equilateral triangle has area 1. Find the total area of all the triangles in the figure below.

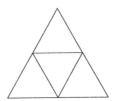

Problem 1.8 How many squares are there in the following diagram?

Problem 1.9 How many angles less than 180° are there in the following diagram?

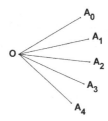

Problem 1.10 In the following diagram, $\angle 1 = 3\angle 3, \angle 2 = 2\angle 3$. The sum of the measures of all the angles is 180°. What is the measure of $\angle AOB$?

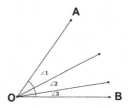

1.3 Practice

Problem 1.11 Count the triangles in each of the following diagrams:

a)

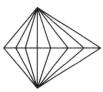
b)

Problem 1.12 Each small equilateral triangle has area 1 in the diagram below. Find the total area of all the triangles.

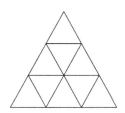

Problem 1.13 How many squares are there in the following diagram?

Problem 1.14 (2008 AMC 8) In the figure, what is the ratio of the area of the gray squares to the area of the white squares?

Problem 1.15 In the following diagram, $ABCD$ is a parallelogram, and each of the segments in the diagram is parallel to one of $\overline{AB}$, $\overline{AD}$, or $\overline{BE}$. Count the number of parallelograms in the diagram that contain the shaded triangle.

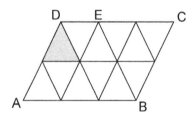

1.3 Practice

Problem 1.16 Arrange several rhombi, all of whose side lengths are 2cm, to form a long parallelogram, as shown in the diagram. Assume the perimeter of the long parallelogram is 144cm, how many rhombi are there?

Problem 1.17 Given a circular disk, use 6 lines to divide the disk into small regions. At most how many regions can there be?

Problem 1.18 (2002 AMC 8) A circle and two distinct lines are drawn on a sheet of paper. What is the largest possible number of points of intersection of these figures?

Problem 1.19 In the diagram, each side is perpendicular to its adjacent sides, and all small sides have equal length. Given that the perimeter of this diagram is 100, find the area of the shape.

Problem 1.20 Use 4 congruent rectangles to form one big square, as shown. The big square has area 100 cm². Suppose the side length of the small square in the center is 2cm. What is the perimeter of each rectangle?

Problem 1.21 (2002 AMC 8) A corner of a tiled floor is shown. If the entire floor is tiled in this way and each of the four corners looks like this one, then what fraction of the tiled floor is made of darker tiles?

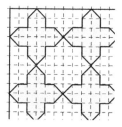

1.3 Practice

Problem 1.22 In the diagram, each side is perpendicular to its adjacent sides, and all small sides have equal length. Given that the area of this shape is 117, find the perimeter.

Problem 1.23 (ZIML 2016) If we cut a circular ring vertically it divides the ring into 2 pieces. A second vertical cut divides the ring into 4 pieces. If we cut the circular ring 10 times vertically, how many pieces do we divide the ring into?

Problem 1.24 Arrange several equilateral triangles and rhombi, all of whose side lengths are 2cm, to form a long parallelogram, as shown in the diagram. Assume the perimeter of the long parallelogram is 244 cm, how many equilateral triangle and rhombi are there?

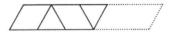

Problem 1.25 (ZIML 2016) How many isosceles triangles can be formed using the dots in the following array as vertices? (The dots are evenly spaced.)

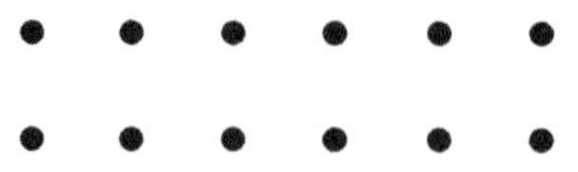

• • • • •

• • • • •

2. Practice with Measurements

In this chapter we cover introductory area questions with a little more depth.

The concepts introduced in this chapter directly correspond to Common Core Math Standards as shown in the following table.

6th Grade	6.RP.1, 6.EE.6
7th Grade	7.NS.1, 7.EE.4, 7.G.1

In addition to the standards above, problems and concepts in this section will help strengthen understanding of the following domains.

6th Grade	6.RP, 6.NS, 6.EE
7th Grade	7.RP, 7.NS, 7.EE, 7.G

Unit Square

A square with dimensions one unit by one unit has area one square unit.

Copyright © ARETEEM INSTITUTE. All rights reserved.

Remark

Common measurements of lengths are inches, centimeters, feet, meters, miles, etc. With these measurements we get areas measured in square inches, square centimeters, square feet, etc.

Example 2.1 Area of a Rectangle

The area of a rectangle with base b and height h is $b \times h$. For example, consider the rectangle in the diagram below,

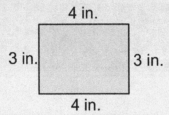

with dimensions 4 inches by 3 inches. The area is thus $3 \times 4 = 12$ square inches. We can understand this answer using unit squares,

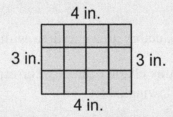

where each of the 12 squares drawn is a unit square, so we see the area is 12 square inches as expected.

Example 2.2 Perimeter and Area of a Rectangle

A rectangle with base b and height h has perimeter $2b + 2h$. For example, suppose we want to find the area of a rectangle that has perimeter 12 and a base that is twice as long as the height. Then we know that half the perimeter $b + h$ is equal to 6. Since the base is twice the height, we see that the base is 2 and the height is 4. This rectangle thus has area $2 \times 4 = 8$.

Copyright © ARETEEM INSTITUTE. All rights reserved.

Coordinate Plane

It is often helpful to label points in a grid. Consider a horizontal number line, called the *x*-axis, combined with a vertical number line, called the *y*-axis as in the diagram below.

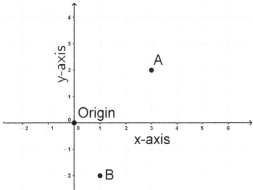

The intersection of the two axes is called the origin. For any point on the plane, its horizontal distance from the origin is called the *x*-coordinate, its vertical distance from the origin is called the *y*-coordinate, and the point is labeled with coordinates (x, y). If the point is to the right of the origin, the *x*-coordinate is positive, otherwise it is negative; similarly, if the point is above the origin, the *y*-coordinate is positive, otherwise it is negative. For example, the point A in the diagram above has coordinates $(3, 2)$ and the point B is $(1, -2)$. The origin has coordinates $(0, 0)$.

Example 2.3

Let A, B, C, D have coordinates $(-1, 1), (-1, 4), (2, 4), (2, 1)$ as in the diagram below.

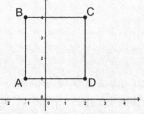

Then $ABCD$ is a square with with area $3 \times 3 = 9$ and perimeter $4 \times 3 = 12$.

2.1 Example Questions

Example 2.4

Two rectangles and two squares are assembled to form a big square as shown. The area of each rectangle is 28 and the area of the small square is 16. What is the area of the entire square?

Solution

The little square's area is 16, so its side length is 4. Since

$$28 = 4 \times 7,$$

we see the rectangles must have dimensions 4 and 7. Therefore the side length of the entire square is

$$4 + 7 = 11$$

and the area of the square is

$$11^2 = 121.$$

Example 2.5

(1999 AMC 8) A rectangular garden 50 feet long and 10 feet wide is enclosed by a fence. To make the garden larger, while using the same fence, its shape is changed to a square. By how many square feet does this enlarge the garden?

Copyright © ARETEEM INSTITUTE. All rights reserved.

2.1 Example Questions

Solution

Note that the perimeter of the rectangular garden is

$$2(50+10) = 120$$

feet. Rearranging 120 feet of fencing into a square yields a square with side length

$$120 \div 4 = 30$$

feet. The area of the rectangular garden is

$$50 \times 10 = 500$$

and the area of the square garden is

$$30^2 = 900,$$

so the area increases by

$$900 - 500 = 400$$

feet.

Example 2.6

A rectangle is divided into 3 squares, as shown in the diagram. Given that the area of one bigger square is 12 in² more than that of one smaller square, find the area of the whole rectangle.

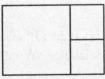

Solution 1

Note the larger square can be divided into four copies of the smaller square.

Hence the area of
$$4 - 1 = 3$$
small squares must be 12, and thus the entire square (equal to six small squares) must be
$$2 \times 12 = 24.$$

Solution 2

Let a be the side length of the bigger square and b the side length of the smaller square, then
$$a = 2b,$$
and
$$a^2 = b^2 + 12.$$
We have $a = 2b$, so
$$a^2 = 4b^2 = b^2 + 12,$$
thus $3b^2 = 12$, and $b^2 = 4$, so
$$a^2 = 16.$$
The area of the rectangle is $a^2 + 2b^2 = 24$ in^2.

Example 2.7

(2000 AMC 8) Figure $ABCD$ is a square. Inside this square three smaller squares are drawn with the side lengths as labeled. What is the area of the shaded L-shaped region?

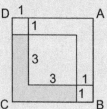

2.1 Example Questions

Solution

Note that the side of the large square has side length

$$1+3+1=5$$

so the area of the large square is $5^2 = 25$. The area of the middle square is 3^2 and the sum of the areas of the two smaller squares is

$$2 \times 1^2 = 2.$$

Therefore, the remaining area of

$$25 - 9 - 2 = 14$$

square units represents the area of the two congruent L-shaped regions. Therefore, the area of one L-shaped region is

$$14 \div 2 = 7.$$

Example 2.8

The shape in the diagram consists of 2 congruent squares and 3 congruent rectangles, and its perimeter is 14. If $AB = 2 \times BC$, what is the area of rectangle $ABCD$?

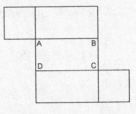

Solution

Since

$$AB = 2 \times BC,$$

the rectangles all have a base that is twice the side length of the squares.

Since the rectangles' heights are the same as the side length of the square, we have that the perimeter of the shape is 14 times the side length of the square.

Hence the side length of the square is 1 so the rectangles have area

$$2 \times 1 = 2.$$

Example 2.9

(2000 AMC 8) The area of rectangle $ABCD$ is 72 units squared. If point A and the midpoints of $\overline{BC}$ and $\overline{CD}$ are joined to form a triangle, what is this triangle's area?

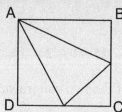

Solution

We are only given that the area of the rectangle is 72 square units, so let's assume that it has base 12 and height 6. If we let M be the right midpoint, and N be the bottom midpoint, then we know that $DN = NC = 6$, and $BM = MC = 3$. The area of triangle ADN is then

$$\frac{1}{2} \times 6 \times 6 = 18.$$

Similarly, the area of triangle MNC is

$$\frac{1}{2} \times 3 \times 6 = 9$$

and the area of triangle ABM is

$$\frac{1}{2} \times 12 \times 3 = 18.$$

Subtracting these from the entire rectangle we get

$$72 - 18 - 9 - 19 = 27$$

units squared as the area of triangle ANM as needed.

Copyright © ARETEEM INSTITUTE. All rights reserved.

2.1 Example Questions

Example 2.10

Divide a big square into 6 congruent rectangles, as shown. Given that each of the rectangles has perimeter 140, find the area of the big square.

Solution

Let a be the length of each of the rectangles, and b be the width. Then the perimeter is 140 so
$$a+b=70,$$
and
$$a=6b.$$
Therefore $7b=70$ so
$$b=70\div 7=10$$
and hence $a = 6 \times 10 = 60$. Therefore the side length of the square is $a = 60$, and the area of the big square is
$$60^2 = 3600.$$

Example 2.11

Four congruent rectangles and one square are assembled into one big square. The areas of the two squares are 144 and 16 respectively. What are the length and width of the rectangles?

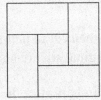

Copyright © ARETEEM INSTITUTE. All rights reserved.

Solution

The side lengths of the squares are 12 and 4 respectively. Note the big square's side length is equal to the small square's side length plus two widths of the rectangles. Hence the rectangles have width

$$(12 - 4) \div 2 = 4.$$

Similarly, the side length is also the width plus the length of the rectangles, so

$$12 - 4 = 8$$

is the length of the rectangles.

Example 2.12

(2001 AMC 8) A square piece of paper, 4 inches on a side, is folded in half vertically. Both layers are then cut in half parallel to the fold. Three new rectangles are formed, a large one and two small ones. What is the ratio of the perimeter of one of the small rectangles to the perimeter of the large rectangle?

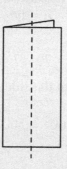

Solution

Since the paper is folded in half and then cut in half the same way, the smaller rectangles each have the same height as the original square and have $\frac{1}{4}$ the length.

Furthermore, since the paper is cut in half after the fold but the fold retains both sides of the larger rectangle, the larger rectangle has the same height as the original square

2.1 Example Questions

but has $\frac{1}{2}$ the length. Therefore, the smaller rectangles have dimensions 4×1, the larger rectangle has dimensions 4×2. Hence the small rectangle has perimeter

$$2(4+1) = 10$$

while the larger has perimeter

$$2(4+2) = 12$$

and the ratio of their perimeters is $10 : 12 = 5 : 6$.

Example 2.13

(2008 AMC 8) Ms.Osborne asks each student in her class to draw a rectangle with integer side lengths and a perimeter of 50 units. All of her students calculate the area of the rectangle they draw. What is the difference between the largest and smallest possible areas of the rectangles?

Solution

Note that the rectangle's area is maximized when its shape most closely resembles a square, or equivalently when the two integer side lengths are closest together. This occurs with the dimensions

$$12 \times 13 = 156.$$

Likewise, the area is smallest when the side lengths have the greatest difference, which occurs with dimensions

$$1 \times 24 = 24.$$

Therefore, the difference in area between the largest and smallest rectangle is $156 - 24 = 132$.

2.2 Quick Response Questions

Problem 2.1 Suppose that the diagram below is not to scale and that the length of AB is 12 and the length of AD is 10. Furthermore, assume that $AD = BE$ and all triangles in the figure below are congruent.

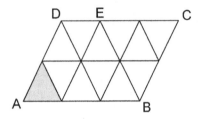

Find the perimeter of parallelogram $ABCD$.

Problem 2.2 Find the perimeter of the one of the small triangles in the above diagram.

Problem 2.3 Suppose I remove the bases of all the small triangles and I want to travel on a zig-zag path from A to B. How long is the path I take?

2.2 Quick Response Questions

Problem 2.4 Suppose that the diagram is not to scale and this time you know that the largest triangle in the diagram (made up of the small triangles) has area 8. (You do not know the lengths *AB* or *CD*.)

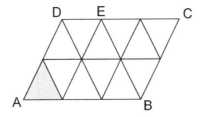

Find the area of a single small triangle.

Problem 2.5 Find the area of the entire parallelogram.

2.3 Practice

Problem 2.6 A rectangle is divided into 3 squares, as shown in the diagram. Given that the area of the rectangle is 150 in^2. Find the length and width of the rectangle.

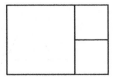

Problem 2.7 A big rectangle is divided into 6 squares of different sizes, as shown. Given that the smallest square in the middle has area 4 cm^2 and the length of the big rectangle is 26, find the area of the big rectangle.

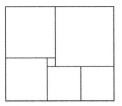

2.3 Practice

Problem 2.8 A big rectangle is divided into 7 smaller congruent rectangles, as shown. Given that the area of the big rectangle is 42 cm^2, find the perimeter of the big rectangle.

Problem 2.9 A rectangle is divided into 4 smaller rectangles by two lines, as shown. The perimeters of three of these rectangles are 12, 14, and 14. Find the perimeter of the remaining (shaded) rectangle.

Problem 2.10 (2008 AMC 8) In square $ABCE$, $AF = 2FE$ and $CD = 2DE$. What is the ratio of the area of $\triangle BFD$ to the area of square $ABCE$?

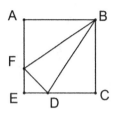

Problem 2.11 The perimeter of rectangle $ABCD$ is 20 cm. Construct a square on the top and right sides of $ABCD$ as shown below. Given that the sum of the areas of these squares is 52 cm^2, find the area of rectangle $ABCD$.

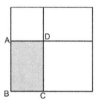

2.3 Practice

Problem 2.12 Two rectangles and one square are assembled to form a big square as shown. The areas of the rectangles are 44 and 28. What is the area of the smaller (lower-right) square?

Problem 2.13 A rectangle is divided into 4 squares, as shown in the diagram. Given that the area of the bigger square is 16 in² more than one of the smaller squares, find the area of the whole rectangle.

Problem 2.14 A rectangle is divided into 4 squares, as shown in the diagram. Given that the area of the rectangle is 300 in^2. Find the length and width of the rectangle.

Problem 2.15 (2009 AMC 8) The length of a rectangle is increased by 10% percent and the width is decreased by 10% percent. What percent of the old area is the new area?

Problem 2.16 A big rectangle is divided into 10 smaller congruent rectangles, as shown. Given that the area of the big rectangle is 120 cm^2, find the perimeter of the big rectangle.

2.3 Practice

Problem 2.17 A rectangle is divided into 4 smaller rectangles by two lines, as shown. The areas of three of these rectangles are 6, 12, and 10. Find the area of the remaining (shaded) rectangle.

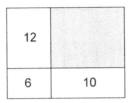

Problem 2.18 (2011 AMC 8) Two congruent squares, $ABCD$ and $PQRS$, have side length 15. They overlap to form the 25 by 15 rectangle $AQRD$ shown. What percent of the area of rectangle $AQRD$ is shaded?

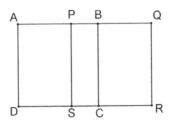

Problem 2.19 The perimeter of rectangle *ABCD* is 10 cm. Construct a square on the top and right sides of *ABCD* as shown below. Given that the sum of the areas of these squares is 13 cm^2, find the area of rectangle *ABCD*.

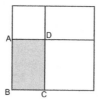

Problem 2.20 The shape in the diagram consists of 2 congruent squares and 3 congruent rectangles, and its perimeter is 28. If $AB = 2 \times BC$, what is the total area of the diagram?

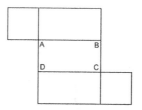

Problem 2.21 (2012 AMC 8) A rectangular photograph is placed in a frame that forms a border two inches wide on all sides of the photograph. The photograph measures 8 inches high and 10 inches wide. What is the area of the border, in square inches?

Problem 2.22 Four congruent rectangles and one square are assembled into one big square. The areas of the two squares are 64 and 16 respectively. What are the length and width of the rectangles?

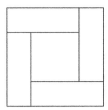

Problem 2.23 Divide a big square into 6 congruent rectangles, as shown. Given that each of the rectangles has perimeter 100, find the area of the big square.

Problem 2.24 (2012 AMC 8) A square with integer side length is cut into 10 squares, all of which have integer side length and at least 8 of which have area 1. What is the smallest possible value of the length of the side of the original square?

Problem 2.25 Given coordinates $A = (0,4)$, $B = (4,4)$, $C = (4,0)$, $D = (1,0)$, $E = (0,0)$, and $F = (0,1)$, what is the ratio of the area of $\triangle BFD$ to the area of square $ABCE$?

3. A Dance with Angles

In this chapter we study angles in detail, learn about the measures of angles, and explore the relations among different angles.

The concepts introduced in this chapter directly correspond to Common Core Math Standards as shown in the following table.

6th Grade	6.EE.6
7th Grade	7.RP.3, 7.EE.4, 7.G.1, 7.G.5
8th Grade	8.EE.7, 8.G.2

In addition to the standards above, problems and concepts in this section will help strengthen understanding of the following domains.

6th Grade	6.NS, 6.EE
7th Grade	7.RP, 7.NS, 7.EE, 7.G
8th Grade	8.EE, 8.G

Copyright © ARETEEM INSTITUTE. All rights reserved.

Lines, Line Segments, and Rays

Recall that a line was an object extending in both directions forever. The part of a line between two points was called a line segment. A ray starts at one point and then follows a line forever in one direction.

Angles Revisited

An angle is formed when two rays have a common origin. This common origin is called the vertex, and the two rays are the sides of the angle. We use the $\angle$ symbol to denote angles. When there is no confusion, we often use the vertex to label an angle. For example, an angle with a vertex of point A might be denoted $\angle A$.

Example 3.1 Measuring Angles

We have already seen that angles are often measured using degrees. A full circle is $360°$. Half of a circle is $180°$.

Perpendicular Lines

Two lines (or rays or segments) that form $90°$ angles are called perpendicular. If the line segments $\overline{AB}$ and $\overline{CD}$ are perpendicular, we write $\overline{AB} \perp \overline{CD}$.

Classifying Angles

- An angle less than 90° is called an acute angle.
- An angle equal to 90° is called a right angle.
- An angle between 90° and 180° is called an obtuse angle.
- An angle equal to 180° is called a straight angle.
- An angle greater than 180° is called a reflex angle.

Example 3.2 Vertical, Adjacent, and Supplementary Angles

Consider the diagram below with angles $\angle 1, \angle 2, \angle 3, \angle 4$.

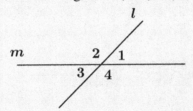

We call angles $\angle 1, \angle 3$ or $\angle 2, \angle 4$ vertical angles. Vertical angles are equal so $\angle 1 = \angle 3$ and $\angle 2 = \angle 4$. We call angles such as $\angle 1, \angle 2$ adjacent angles. These angles add up to 180°, so $\angle 1 + \angle 2 = 180°$. In general, we call angles that add up to 180° supplementary angles.

Remark

Two angles that add up to 90° are called complementary angles.

Parallel Lines and Transversals

Recall two lines that do not intersect (lines *m* and *n* in the diagram) are called parallel. We use the notation $m \parallel n$ to denote the m, n are parallel.

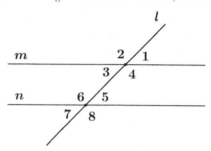

Line *l* intersecting both parallel lines is referred to as a transversal.

Example 3.3 Angles in Parallel Lines

In the above diagram, we have
- Angles such as $\angle 1, \angle 5$ or $\angle 4, \angle 8$ are called corresponding angles. Corresponding angles are equal so $\angle 1 = \angle 5$ and $\angle 4 = \angle 8$.
- Angles such as $\angle 3, \angle 5$ or $\angle 4, \angle 6$ are called alternating interior angles. Alternating interior angles are equal so $\angle 3 = \angle 5$ and $\angle 4 = \angle 6$.
- Angles such as $\angle 1, \angle 7$ or $\angle 2, \angle 8$ are called alternating exterior angles. Alternating exterior angles are equal so $\angle 1 = \angle 7$ and $\angle 2 = \angle 8$.
- Angles such as $\angle 3, \angle 6$ or $\angle 4, \angle 5$ are called same-side interior angles. Alternating interior angles are supplementary so $\angle 3 + \angle 6 = 180°$ and $\angle 4 + \angle 5 = 180°$.
- Angles such as $\angle 1, \angle 8$ or $\angle 2, \angle 7$ are called same-side exterior angles. Alternating exterior angles are supplementary so $\angle 1 + \angle 8 = 180°$ and $\angle 2 + \angle 7 = 180°$.

Remark

The above are also helpful in proving lines are parallel. For example, if a pair of corresponding angles are equal, then the line must be parallel. This fact holds for the other types of angles as well.

Copyright © ARETEEM INSTITUTE. All rights reserved.

3.1 Example Questions

Example 3.4

Consider the following diagram of two parallel lines and two transversals which meet at point *C*.

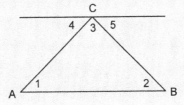

Which angles are equal in the above figure? Why?

Solution

Since the two lines are parallel, by alternating interior angles, $\angle 1 = \angle 4$ and $\angle 2 = \angle 5$.

Example 3.5

Consider the previous diagram of two parallel lines and two transversals which meet at point *C*. What is $\angle 3 + \angle 4 + \angle 5$?

Solution

Note that combining $\angle 3$, $\angle 4$, and $\angle 5$ gives a straight line. Hence
$$\angle 3 + \angle 4 + \angle 5 = 180°.$$

Example 3.6

Consider the previous diagram of two parallel lines and two transversals which meet at point *C*. What is the sum of the angles in triangle *ABC*? Why?

Copyright © ARETEEM INSTITUTE. All rights reserved.

Solution

By the previous problems, we know

$$\angle 3 + \angle 4 + \angle 5 = 180°$$

and

$$\angle 1 = \angle 4$$

and

$$\angle 2 = \angle 5.$$

Therefore, subsittuting we get

$$\angle 1 + \angle 2 + \angle 3 = 180°.$$

Remark

The sum of the three angles in any triangle is 180°. This is one of the first theorems in Euclidean geometry. The example above shows one of the proofs of this theorem.

Example 3.7

(1999 AMC 8) What is the degree measurement of angle A?

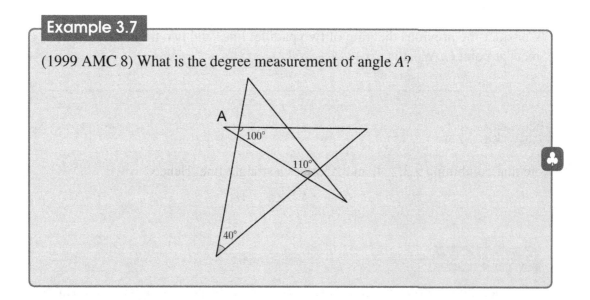

3.1 Example Questions

Solution

The supplementary angle of 110° is 70°. Therefore, the bottom triangle contains angles of 40° and 70° so has a third angle with measure

$$180° - 70° - 40° = 70°.$$

Now look at the small triangle containing angle *A*. Using vertical angles the bottom angle is 70°. Using supplementary angles the top right angle is

$$180° - 100° = 80°.$$

Thus angle *A* is

$$180° - 70° - 80° = 30°.$$

Example 3.8

Suppose that in the diagram below we have parallel lines and a transversal.

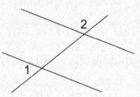

If the measure of ∠1 is half the measure of ∠2, find the measure of ∠1.

Solution

Angles ∠1, ∠2 are same-side exterior angles, hence they are supplementary. As

$$\angle 2 = 2 \times \angle 1,$$

we have

$$\angle 1 + \angle 2 = 3 \times \angle 1 = 180°$$

and hence

$$\angle 1 = 180° \div 3 = 60°.$$

Copyright © ARETEEM INSTITUTE. All rights reserved.

Example 3.9

Consider the diagram below:

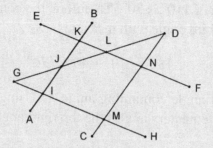

Suppose we know that $\overleftrightarrow{AB}$ and $\overleftrightarrow{CD}$ are parallel, $\angle DMH = 75°$, $\angle ELD = 140°$, and $\angle AJG = 35°$. What is $\angle LND$?

Solution

We first have
$$\angle DLN = 180° - \angle ELD = 180° - 140° = 40°$$
because adjacent angles are supplementary. Since $\overline{AB} \parallel \overline{CD}$,
$$\angle NDL = \angle AJG = 35°$$
as they are corresponding angles. Therefore, as the angles in triangle $\triangle DNL$ add up to $180°$,
$$\angle LND = 180° - 40° - 35° = 105°.$$

3.1 Example Questions

Example 3.10

Consider the following "star" diagram, not drawn to scale.

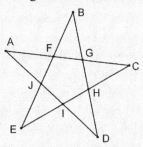

Suppose $\angle EFC = 120°$, $\angle AIC = 100°$, and $\angle BEC = 35°$. Calculate $\angle CAD$.

Solution

Using supplementary angles we know

$$\angle AFJ = 180° - 120° = 60°$$

and similarly

$$\angle AIE = 180° - 100° = 80°.$$

Using $\triangle JEI$ we calculate

$$\angle EJI = 180° - 35° - 80° = 65°.$$

Using vertical angles $\angle AJF = \angle EJI = 65°$. Hence

$$\angle CAD = \angle JAF = 180° - 65° - 60° = 55°.$$

Example 3.11

Given the figure below, if $\angle DFE = 60°$ and $\angle BCF = 90°$, what is the measure of $\angle CAF$?

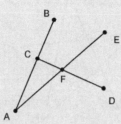

Solution

Given $\angle DFE = 60°$, we observe that by vertical angles,

$$\angle AFC = 60°.$$

Given $\angle BCF = 90°$, we get

$$\angle FCA = 180° - 90° = 90°.$$

Therefore,
$$\angle AFC = 180° - 90° - 60° = 30°$$

since the angles in triangle AFC add up to $180°$.

Example 3.12

Suppose $\angle AOF = 180°$ and is divided into five equal angles as shown below.

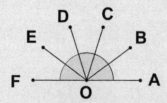

If $\angle AOB = \angle BOC = \angle COD = angle DOE = \angle EOF$, find $\angle BOE$.

3.1 Example Questions

Solution

All five angles are equal so each angle is

$$180 \div 5 = 36°.$$

Since $\angle BOE$ contains three of the smaller angles,

$$\angle BOE = 3 \times 36° = 108°.$$

Example 3.13

(2000 AMC 8) In triangle CAT, we have $\angle ACT = \angle ATC$ and $\angle CAT = 36°$. If $\overline{TR}$ bisects $\angle ATC$, then what is the measure of $\angle CRT$?

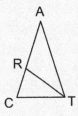

Solution

Given $\triangle ACT$, we have that the three angles add up to $180°$ and $\angle C = \angle T$ (since $\triangle ACT$ is isosceles). Therefore,

$$36° + \angle ATC + \angle ATC = 180°$$

implies that twice $2\angle ATC = 180° - 36° = 144°$ so

$$\angle ATC = 144° \div 2 = 72°.$$

Since $\angle ATC$ is bisected by $\overline{TR}$,

$$\angle RTC = 72 \div 2 = 36.$$

In smaller $\triangle RTC$, the sum of the angles in that triangle is also $180°$, so

$$36° + \angle ATC + \angle CRT = 180°.$$

The same calculation from above gives us that $\angle CRT = 72°$.

3.2 Quick Response Questions

Problem 3.1 Classify each of the following angles and give an estimate of the angle in degrees.

(a)

(b)

(c)

3.2 Quick Response Questions

(d)

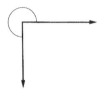

(e)

(f)

Problem 3.2 Let *m* and *n* be a pair of parallel lines and let transversal *l* cut across the parallel lines as shown in the figure below, which is used in the next four questions.

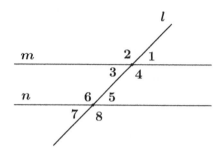

List all corresponding angles in the figure above. List all vertical angles in the figure above.

Problem 3.3 Let's look at alternating angles!

(a) List all alternating interior angles in the figure above.

(b) List all alternating exterior angles in the figure above.

Problem 3.4 Let's look at same-side angles!

(a) List all same-side interior angles in the figure above.

3.2 Quick Response Questions

(b) List all same-side exterior angles in the figure above.

Problem 3.5 In the diagram above, if $\angle 1 = 42°$, find the measures of the other angles.

3.3 Practice

Problem 3.6 Consider the diagram below:

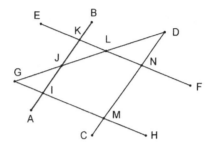

Suppose we know that $\overleftrightarrow{AB}$ and $\overleftrightarrow{CD}$ are parallel, $\angle DMH = 70°$, $\angle ELD = 135°$, and $\angle AJG = 30°$. What is $\angle LND$?

Problem 3.7 Consider the following "star" diagram, not drawn to scale.

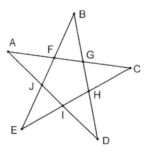

Suppose $\angle EFC = 115°$, $\angle AIC = 95°$, and $\angle BEC = 30°$. Calculate $\angle CAD$.

3.3 Practice

Problem 3.8 Suppose that in the diagram below we have parallel lines and a transversal.

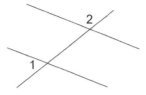

If the measure of ∠1 is one-third the measure of ∠2, find the measure of ∠1.

Problem 3.9 Consider the diagram below:

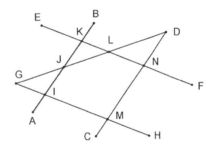

Suppose we know that $\overleftrightarrow{AB}$ and $\overleftrightarrow{CD}$ are parallel, $\angle DMH = 75°$, $\angle ELD = 140°$, and $\angle AJG = 35°$. What is $\angle BKE$?

Problem 3.10 Consider the following "star" diagram, not drawn to scale.

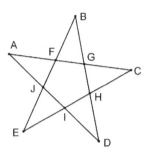

Suppose $\angle EFC = 120°$, $\angle AIC = 100°$, and $\angle BEC = 35°$. Calculate $\angle FJI$.

Problem 3.11 (2000 AMC 8) If $\angle A = 20°$ and $\angle AFG = \angle AGF$, then what is $\angle B + \angle D$?

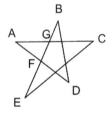

3.3 Practice

Problem 3.12 Given the figure below, if $\angle DFE = 75°$ and $\angle BCF = 95°$, what is the measure of $\angle CAF$?

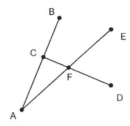

Problem 3.13 Suppose $\angle AOF = 180°$ and is divided into five equal angles as shown below.

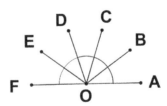

If $\angle AOB = \angle BOC = \angle COD = \angle DOE = \angle EOF$, find $\angle AOE$.

Problem 3.14 Using the above diagram, what is $\angle BOD$?

Problem 3.15 Consider the diagram below, where *l* and *m* are parallel but the drawing is not necessarily to scale.

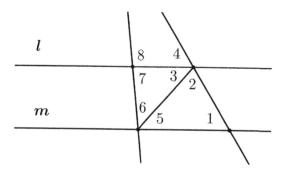

Suppose that ∠1 = 60°, ∠5 = 50°, ∠8 = 105°. Find the measure of ∠6.

Problem 3.16 Using the above diagram with the given measurements, find the measure of ∠3.

Problem 3.17 Suppose you have △ABC with angles ∠A, ∠B, ∠C. If ∠A and ∠C are complementary and ∠B is three times ∠A, what is ∠C?

3.3 Practice

Problem 3.18 Consider the diagram below, where l and m are parallel but the drawing is not necessarily to scale.

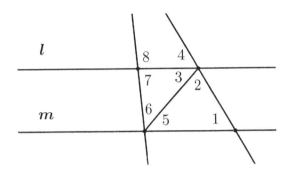

Suppose that $\angle 1 = 60°, \angle 5 = 50°, \angle 8 = 105°$. Find the measure of $\angle 4$.

Problem 3.19 Using the above diagram with the given measurements, find the measure of $\angle 2$.

Problem 3.20 (2014 AMC 8) In $\triangle ABC$, D is a point on side $\overline{AC}$ such that $\angle BCD = \angle CBD = 70°$. What is the degree measure of $\angle ADB$?

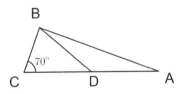

Problem 3.21 Suppose we have the following diagram for the next three problems.

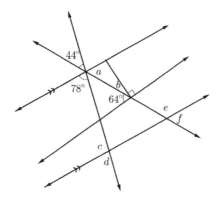

Find angles a and b.

Problem 3.22 Find angles c and d.

Problem 3.23 Find angles e and f.

Problem 3.24 The measures of the three angles of a triangle form an arithmetic sequence, If the smallest angle has measure $33°$, what is the angle measure of the largest angle?

3.3 Practice

Problem 3.25 (MATHCOUNTS 2003) Two complementary angles, A and B, have measures in the ratio of 7 to 23, respectively. What is the ratio of the measure of the complement of angle A to the measure of the complement of angle B? Express your answer as a common fraction.

4. Magic in Triangles

Triangles are the simplest shape among polygons. Studying the properties of triangles can make it easier to understand other shapes, such as quadrilaterals, pentagons, polygons with more sides, and circles. Furthermore, many facts about triangles and problem-solving techniques involving triangles are already quite fascinating.

The concepts introduced in this chapter directly correspond to Common Core Math Standards as shown in the following table.

6th Grade	6.EE.6, 6.G.1
7th Grade	7.RP.2, 7.EE.4, 7.G.1, 7.G.4
8th Grade	8.EE.7, 8.G.8

In addition to the standards above, problems and concepts in this section will help strengthen understanding of the following domains.

6th Grade	6.RP, 6.NS, 6.EE, 6.G
7th Grade	7.RP, 7.NS, 7.EE, 7.G
8th Grade	8.EE, 8.G

Copyright © ARETEEM INSTITUTE. All rights reserved.

Classifying Triangles by Angles

- If a triangle contains an obtuse angle we call it an obtuse triangle.
- If a triangle contains a right angle we call it a right triangle.
- If a triangle only contains acute angles we call it an acute triangle.

Example 4.1

Recall from the previous chapter that angles in a triangle add up to $180°$. Therefore an obtuse triangle has one obtuse angle and two acute angles, a right triangle has one right angle and two complementary acute angles, and an acute triangle has three acute angles.

Classifying Triangles by Sides

- If a triangle has three equal sides we call it an equilateral triangle.
- If a triangle has two equal sides we call it an isosceles triangle.
- If all three sides of a triangle are different we call it a scalene triangle.

Example 4.2 Sides and Opposite Angles

If two sides in a triangle are equal, then the angles opposite those sides are equal. Hence, an equilateral triangle has three equal angles (each $60°$) and an isosceles triangle has 2 equal angles.

Congruent Triangles

Recall that two triangles are congruent if they are exactly the same. That is, they have have the same sides and the same angles.

Copyright © ARETEEM INSTITUTE. All rights reserved.

Similar Triangles

Two triangles are similar if they are the same, except for possibly their size. That is, their angles are equal and all their sides are in a common ratio (for example, each of the sides of one triangle could be twice the corresponding side of the other triangle).

Example 4.3

In the diagram below $\triangle ABC$ is congruent to $\triangle DEF$ (written $\triangle ABC \cong \triangle DEF$). Triangle $\triangle GHI$ is NOT congruent to either $\triangle ABC$ or $\triangle DEF$. However, all three triangles are similar (written $\triangle ABC \sim \triangle DEF \sim \triangle GHI$).

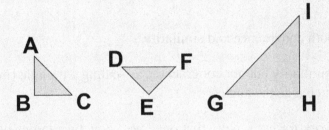

Example 4.4 SSS Rule

- If two triangles have all three pairs of sides the same lengths, then they are actually congruent. Think of making a triangle from three sticks, there is only one triangle you can make! This rule is often referred to as the SSS rule for congruence.
- If two triangles are different sizes, but all the sides share a common ratio to each other they are similar. This rule is referred to as the SSS rule for similarity.

Copyright © ARETEEM INSTITUTE. All rights reserved.

4.1 Example Questions

Example 4.5

For each of the following "rules", state whether they work for proving congruence, similarity, both, or neither. If the rule does not work, give a counterexample.
  (a) SAS (two sides and the angle between them)
  (b) AAA (all three angles)
  (c) ASA (two angles and the side between them)
  (d) AAS (two angles and a side not between them)
  (e) SSA (two sides and an angle not between them)

Solution

SAS works for both congruence and similarity.

AAA works for similarity but not congruence, as scaling a triangle (making it larger or smaller) does not change the angles.

ASA and AAS work for both. Note that once we know two angles we know the third, so these could be referred to as AAA plus S. We already know the triangles are similar (just from AAA), and knowing a side ensures congruence as well.

SSA works for neither. Consider, for example, cutting an isosceles triangle into two unequal pieces (through the third vertex).

Example 4.6

Suppose $\triangle ABC$ is an isosceles triangle with $\angle A = \angle B$. Let D be a point on $\overline{AB}$. Prove that D is the midpoint of $\overline{AB}$ (we call $\overline{CD}$ the *median* from C) if and only if $\angle ACD = \angle BCD$ (we call $\overline{CD}$ the *angle bisector* of $\angle ACB$). Further prove that $\overline{CD}$ is perpendicular to $\overline{AB}$ (we call $\overline{CD}$ the *altitude* from C).

4.1 Example Questions

Solution

Remember we have two directions to prove for the first statement!

First suppose D is the midpoint. We want to show

$$\angle ACD = \angle BCD.$$

Since the triangle is isosceles,
$$AC = BC.$$

By the definition of midpoint,
$$AD = BD.$$

As $CD = CD$, we can use SSS to show

$$\triangle ACD \cong \triangle BCD.$$

Hence
$$\angle ACD = \angle BCD.$$

Now suppose

$$\angle ACD = \angle BCD$$

and we want to show D is the midpoint of $\overline{AB}$. Identical to above we have

$$AC = BC$$

and
$$CD = CD.$$

Therefore, we can use SAS to show

$$\triangle ACD \cong \triangle BCD.$$

Hence $AD = BD$ so D is the midpoint.

Lastly, $\angle ACD$ and $\angle BCD$ are adjacent, so

$$\angle ACD + \angle BCD = 180°.$$

As they are equal, both must be $90°$ and hence $\overline{CD}$ is perpendicular to $\overline{AB}$ as needed.

Copyright © ARETEEM INSTITUTE. All rights reserved.

Example 4.7

Let $\overline{AD}$ and $\overline{BE}$ be medians in $\triangle ABC$. Prove that EF is half of AB.

Solution

First note that $\triangle DCE$ and $\triangle BCA$ share $\angle C$. Further, using the definition of midpoints we have
$$CB = 2 \times CD$$
and
$$CA = 2 \times CE.$$
Hence, $\triangle DCE \sim \triangle BCA$ using SAS. As all the ratios in similar triangles are the same,
$$AB = 2 \times ED$$
as needed.

Example 4.8

Prove that the diagonals in a square are perpendicular.

Solution

Consider the labeling below:

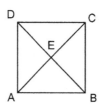

4.1 Example Questions

As a square is made up of four equal sides and four equal angles, using SAS we have
$$\triangle ABC \cong \triangle BCD \cong \triangle CDA \cong \triangle DAB.$$

Further, all of these triangles are isosceles, and hence must have a 90° angle and two $90 \div 2 = 45°$ angles. Hence,
$$\angle ABE = \angle BAE = 45°,$$
and so looking at triangle $\triangle ABE$, we see
$$\angle AEB = 180° - 45° - 45°,$$
so the diagonals are perpendicular as needed.

Example 4.9

Prove that in a quadrilateral $ABCD$, if $\overline{AB}$ is parallel to $\overline{CD}$ and $\overline{BC}$ is parallel to $\overline{AD}$, then $AB = CD$ (and similarly $BC = AD$).

Solution

Draw diagonal BD. Using alternate interior angles, we have
$$\angle ABD = \angle BDC,$$
and similarly
$$\angle ADB = \angle CBD.$$
Hence by ASA
$$\triangle BDA \cong \triangle DBC$$
and $AB = CD$ as needed. The proof that $BC = AD$ is identical.

Example 4.10

Let D, E be midpoints of $\overline{AB}$ and $\overline{BC}$ in $\triangle ABC$. Consider the lines through D and E that are perpendicular to $\overline{AB}$ and $\overline{BC}$ respectively. These lines are called the *perpendicular bisectors* of $\overline{AB}$ and $\overline{BC}$. Let F be the intersection of these perpendicular bisectors. Show that AF, BF, CF all have the same length.

Solution

Since D and E are midpoints we know $AD = BD$ and $BE = CE$. Thus by SAS we have

$$\triangle ADF \cong \triangle BDF$$

and similarly

$$\triangle BEF \cong CEF.$$

Therefore we have $AF = BF$ and $BF = CF$ so all three have the same length.

> **Example 4.11**
>
> Answer the following questions, with explanations.
> (a) Is it true that any three points are the vertices of some triangle?
> (b) Given three numbers $p \leq q \leq r$ such that $p + q + r = 180$, is there a triangle $\triangle ABC$ such that $\angle A = p, \angle B = q, \angle C = r$?
> (c) Given three numbers $a \leq b \leq c$, is there a triangle $\triangle ABC$ with side lengths a, b, c? What can go wrong? Can you come up with a rule for when you can create a triangle with sides lengths a, b, c?

Solution

For (a), if all three points are in a straight line, we do not get a triangle. Note: this is sometimes called a *degenerate* triangle. However, if they are not all in a straight line, there is exactly one triangle with the three points as vertices.

For (b), yes, there are many different triangles. Start with a base $\overline{AB}$ and draw angles of $p°$ and $q°$ as in the diagram below:

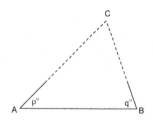

4.1 Example Questions

Let C be the intersection of the rays from the two angles. Note then in $\triangle ABC$,

$$\angle A = p, \angle B = q, \angle C = 180° - p - q = r$$

as needed. Further, if we start with a different size for the base, we get a different (non-congruent) triangle.

For (c), we just need to ensure that the two shortest sides are long enough to reach the third. As long as they are long enough, we can put the sides together to make a triangle. Hence the rule is

$$c < a + b.$$

Note this rule is referred to as the *triangle inequality*.

Example 4.12

Prove that if you connect the midpoints of the sides of an equilateral triangle it divides the triangle into four smaller congruent equilateral triangles.

Solution

Since all the side lengths of the triangle are equal, we can use SAS to show the three outer triangles are all congruent.

Further, as they are isosceles with a 60° angle, they in fact are all equilateral triangles.

Lastly, SSS shows that the inner triangle is congruent to the outer ones.

Thus all four triangles are equilateral triangles and all congruent.

Example 4.13

Recall the "star" diagram from earlier (again do not assume it is drawn to scale).

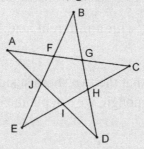

Now suppose that $\angle A = 40°$ and that $\triangle AFJ$ is isosceles. Calculate $\angle B + \angle D$.

Solution

We know $\triangle AFJ$ is isosceles, so
$$\angle AFJ = \angle AJF = (180° - 40°) \div 2 = 70°,$$
as angles add up to $180°$ in a triangle. Using adjacent angles,
$$\angle FJI = 180° - 70° = 110°.$$
Noting that the angles in $\triangle BJD$ also add up to $180°$, we have
$$\angle B + \angle D = 180° - 110° = 70°.$$

Example 4.14

(2002 AMC 8) The area of triangle XYZ is 8 square inches. Points A and B are midpoints of congruent segments $\overline{XY}$ and $\overline{XZ}$. Altitude $\overline{XC}$ bisects $\overline{YZ}$.

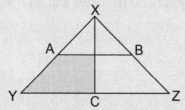

What is the area (in square inches) of the shaded region?

Copyright © ARETEEM INSTITUTE. All rights reserved.

Solution

Consider the diagram below

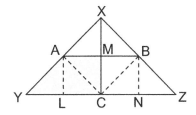

where $\triangle XYZ$ is divided into 8 triangles and $\overline{AL}$ and $\overline{BN}$ are also altitudes.

Arguments very similar to the ones done earlier in this chapter show that

$$\triangle AXM, \triangle BXM, \triangle YAL, \triangle CAL, \triangle ACM, \triangle BCM, \triangle CBN, \triangle ZBN$$

are all congruent.

Therefore each of the 8 triangles has area

$$8 \div 8 = 1$$

square inch, so the shaded region (which is made up of 3 of these triangles) has area 3 square inches.

4.2 Quick Response Questions

Problem 4.1   (a) Draw 2 different equilateral triangles.

(b) Draw 2 different isosceles triangles.

(c) Draw 2 different scalene triangles.

(d) Draw 2 different obtuse triangles.

(e) Draw 2 different right triangles.

(f) Draw 2 different acute triangles.

4.2 Quick Response Questions

Problem 4.2 State whether you think each of the triangles below is acute, obtuse, or right.

(a)

(b)

(c)

(d)

(e)

(f)

4.2 Quick Response Questions

Problem 4.3 Consider the diagram below:

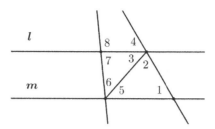

Suppose $\angle 1 = 45°, \angle 2 = 75°, \angle 6 = 60°, \angle 7 = 70°$. Are l and m parallel?

Problem 4.4 Prove that alternate interior angles are equal and that same-side exterior angles are supplementary using the earlier results proved in class. That is, prove (for example) that $\angle 3 = \angle 5$ and that $\angle 1 + \angle 8 = 180°$ in the diagram below. Try to only use facts about vertical angles, adjacent angles, and corresponding angles in your proof.

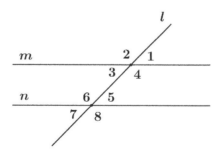

Problem 4.5 Suppose a triangle has a 60° angle and the two sides adjacent to it are equal. Is it an equilateral triangle?

4.3 Practice

Problem 4.6 Prove that in a triangle an exterior angle is equal to the sum of the two interior angles not adjacent to it.

Problem 4.7 Prove that two angles are equal in a triangle if and only if the opposite sides are equal. (Recall that to prove an "if and only if" statement you need to prove both directions!)

Problem 4.8 Suppose an isosceles triangle has an angle of 30°. Find all possibilities for the remaining two angles.

Problem 4.9 Consider the diagram below.

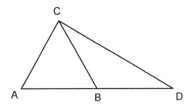

Suppose $AB = AC = 1$ and $\angle BAC = 60°$ and $\angle ADC = 30°$. Find BD.

4.3 Practice

Problem 4.10 Prove that if you connect the midpoints of the sides of a triangle it divides the triangle into four smaller congruent triangles.

Problem 4.11 Prove that the diagonals of a parallelogram bisect each other.

Problem 4.12 Prove that the perpendicular bisectors of the three sides of a triangle $\triangle ABC$ all meet in a single point.

Problem 4.13 For each of the following, suppose you have $\triangle ABC$ satisfying the information given. Calculate as much of the missing information as you can. It may help to draw a diagram.

(a) $a = b = c = 2$.

(b) $a = b = 4$ and $\angle C = 40°$.

(c) $\angle A = 50°, \angle C = 80°, a = 10$.

(d) $\angle A = 40°, \angle B = 70°, a = 10$.

Problem 4.14 Suppose the two heights outside an obtuse triangle are the same length. Prove that the triangle is isosceles.

Problem 4.15 Suppose we have a "star" diagram as below (do not assume it is drawn to scale).

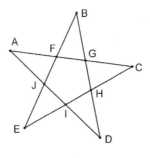

Now suppose that $\angle A = 30°$ and that $\triangle AFJ$ is isosceles. Calculate $\angle B + \angle D$.

4.3 Practice

Problem 4.16 (2006 AMC 8) Triangle *ABC* is an isosceles triangle with $\overline{AB} = \overline{BC}$. Point *D* is the midpoint of both $\overline{BC}$ and $\overline{AE}$, and $\overline{CE}$ is 11 units long. Triangle *ABD* is congruent to triangle *ECD*.

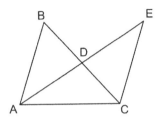

What is the length of $\overline{BD}$?

Problem 4.17 Suppose that *ABCD* is a square. Let point *E* be *outside* the square and that △*CDE* is an equilateral triangle (see the diagram). What is the measure of ∠*EAD*?

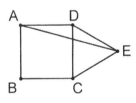

Problem 4.18 Given square *ABCD*, let *P* and *Q* be the points outside the square that make triangles *CDP* and *BCQ* equilateral. Segments $\overline{AQ}$ and $\overline{BP}$ intersect at *G*. Find angle *AGP*.

Problem 4.19 Given square $ABCD$, let P and Q be the points outside the square that make triangles CDP and BCQ equilateral. Prove that triangle APQ is also equilateral.

Problem 4.20 Suppose in the diagram below that $\triangle ABC$ is isosceles and $\angle CAG = 20°$. Find the measure of $\angle ABD$.

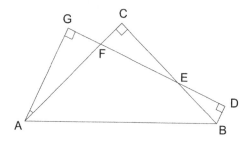

Problem 4.21 (2015 AMC 8) What is the smallest whole number larger than the perimeter of any triangle with a side of length 5 and a side of length 19?

Problem 4.22 Suppose triangle ABC is formed by the coordinates $A = (0,0)$, $B = (2,0)$, and $C = (0,4)$. A second congruent triangle DEF is formed with the two given points $D = (0,0)$ and $E = (0,2)$. If at least one of the coordinates of F is negative, what are the possible coordinates of point F?

Copyright © ARETEEM INSTITUTE. All rights reserved.

4.3 Practice

Problem 4.23 Suppose that $ABCD$ is a square. Let point E be *inside* the square and that $\triangle CDE$ is an equilateral triangle. What is the measure of $\angle EAD$?

Problem 4.24 (ZIML 2016) In $\triangle ABC$, $AB = AC$. Point D is on side $\overline{AB}$ such that $\overline{CD}$ bisects $\angle ACB$, and $CD = BC$. Find the measure of $\angle BAC$ in degrees.

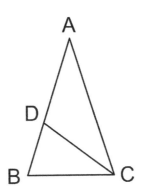

Problem 4.25 (2005 AMC 8) How many different isosceles triangles have integer side lengths and perimeter 23?

5. You Are Special, Right?

In this chapter, we focus on right triangles, in particular, right triangles with some special angles.

The concepts introduced in this chapter directly correspond to Common Core Math Standards as shown in the following table.

6th Grade	6.RP.1, 6.RP.3, 6.EE.1, 6.EE.6, 6.G.1
7th Grade	7.EE.4, 7.G.1, 7.G.5, 7.G.6
8th Grade	8.EE.2, 8.G.6, 8.G.7, 8.G.8

In addition to the standards above, problems and concepts in this section will help strengthen understanding of the following domains.

6th Grade	6.RP, 6.EE, 6.G
7th Grade	7.RP, 7.EE, 7.G
8th Grade	8.EE, 8.G

Copyright © ARETEEM INSTITUTE. All rights reserved.

Example 5.1 Labeling a Triangle

Often, in a triangle $\triangle ABC$ we will label the angles as $\angle A, \angle B, \angle C$. Since the angles are often related to the size of the opposite sides, we often label the sides $AB = c$, $AC = b$, and $BC = a$ as in the diagram below:

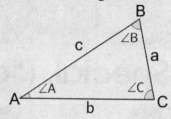

Pythagorean Theorem

In a right triangle $\triangle ABC$ with $\angle C = 90°$ we have $c^2 = a^2 + b^2$ (a, b, c are defined as above).

Remark

On the other hand, if we know that $a^2 + b^2 = c^2$ in a triangle with sides a, b, c, then the triangle must be a right triangle. This is called the converse to the Pythagorean theorem.

Example 5.2 Distance Formula

Suppose A has coordinates (x_1, y_1) and B has coordinates (x_2, y_2). Then if we let point C have coordinates (x_2, y_1) we have a right triangle $\triangle ABC$ with legs of length $x_2 - x_1$ and $y_2 - y_1$ as in the diagram below.

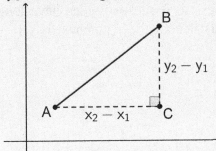

Thus, the hypotenuse has length $\sqrt{(x_2 - x_1)^2 + (y_2 - y_1)^2}$ which is the distance from point A to point B.

Special Angles

In a triangle we call the angles $30°, 45°, 60°, 90°$ special angles, because (as we will see) they have many nice properties.

Example 5.3 45-45-90 Triangle

Suppose you have an isosceles right triangle. Since angles in a triangle add up to 180°, the two equal angles must add up to 90°, so they are each 45°. We call this triangle a 45-45-90 triangle.

The triangle is isosceles, so if the two sides have length 1, we can use the Pythagorean theorem to calculate the hypotenuse is $\sqrt{1^2 + 1^2} = \sqrt{2}$.

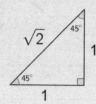

In general, all 45-45-90 triangles are similar, so the sides of any 45-45-90 triangle must be in the ratio $1 : 1 : \sqrt{2}$.

Example 5.4 Equilateral Triangle

Recall that an equilateral triangle has three equal sides and three equal angles. Since angles in a triangle must add up to 180°, we know that each of the angles is 60°.

Example 5.5 30-60-90 Triangle

Suppose we start with a equilateral triangle with side length 2. If we then cut the triangle into two congruent pieces as in the diagram we get two 30-60-90 triangles.

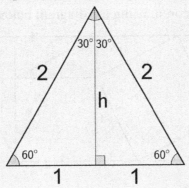

Since we have right triangles, we can use the Pythagorean theorem to say $1^2 + h^2 = 2^2$ so we can find the height $h = \sqrt{3}$. That is, we get the 30-60-90 triangle shown below with sides $1, \sqrt{3}, 2$ opposite from the angles $30°, 60°, 90°$.

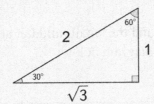

In general, all 30-60-90 triangles are similar, so the sides of any 30-60-90 triangle must be in the ratio $1 : \sqrt{3} : 2$.

5.1 Example Questions

Example 5.6

Prove the Pythagorean theorem using the diagram below:

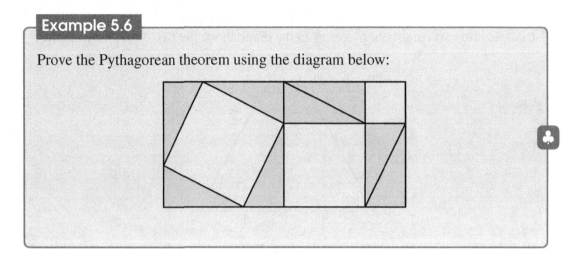

Solution

Let the triangles be denoted T, and the small, middle, and big squares denoted (respectively) S, M, B as labeled in the diagram below:

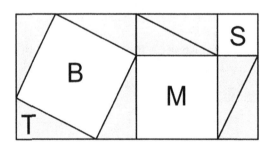

Note the entire diagram is made up of two congruent squares, so they have same area. The left square is
$$B + 4T$$
and the right square is
$$S + M + 4T.$$

Since the 4 triangles on each side have the same area, the area of square B is the sum of the squares M, S. If T has side lengths $a < b < c$, it is easy to see the areas of S, M, L are

5.1 Example Questions

respectively a^2, b^2, c^2 and thus
$$a^2 + b^2 = c^2$$
as needed.

Example 5.7

Prove the converse to the Pythagorean theorem.

Solution

Suppose we have $\triangle ABC$ with
$$AB = c, AC = b, BC = a$$
and
$$c^2 = a^2 + b^2.$$
We want to show that $\triangle ABC$ is a right triangle.

Form a new right triangle $\triangle A'B'C'$ with $A'C' = b$, $B'C' = a$, and $\angle C = 90°$. By the Pythagorean theorem,
$$A'B' = c^2.$$
Hence, using SSS, $\triangle ABC \cong \triangle A'B'C'$ and therefore $\angle C' = 90°$. Thus $\triangle ABC$ is a right triangle as needed.

Example 5.8

For each of the following, state whether it is possible to have a right triangle with the given side lengths. If it is possible, we call (a,b,c) a *Pythagorean Triple*.
   (a) $3, 4, 5$.
   (b) $4, 5, 6$.
   (c) $5, 12, 13$.
   (d) $6, 8, 10$.
   (e) $5, 7, 8$.

Solution

By the converse to the Pythagorean theorem, a triangle with sides a,b,c is right if $a^2 + b^2 = c^2$. Checking we see

$$(3,4,5), (5,12,13), (6,8,10)$$

are Pythagorean triples, while the others are not.

Example 5.9

Suppose you have a right triangle $\triangle ABC$ with hypotenuse $AC = 13$. Attach right triangle $\triangle BCD$ with hypotenuse BC to the side of $\triangle ABC$. If $\triangle BCD$ has sides of length $3, 4$, find AB.

Solution

Using the Pythagorean theorem

$$BC^2 = BD^2 + BC^2,$$

so $BC = 5$ (recall $(3,4,5)$ is a Pythagorean triple). Using the Pythagorean theorem again

$$AC^2 = AB^2 + BC^2,$$

so $AB = 12$ (as $(5,12,13)$ is another Pythagorean triple!).

Example 5.10

Let $ABCD$ be a rectangle with $AB = 6, BC = 3$. Let E be the point a third of the way from A to B on $\overline{AB}$. Is $\angle CED$ a right angle?

Copyright © ARETEEM INSTITUTE. All rights reserved.

5.1 Example Questions

Solution

Since E is a third of the way from A to B, we have
$$AE = 6 \div 3 = 2$$
and
$$BE = 2 \times 2 = 4.$$
Therefore, we can use the Pythagorean theorem on right triangles $\triangle BCE$ and $\triangle ADE$ to get that
$$CE^2 = 25, DE^2 = 13.$$
We then see
$$36 = DC^2 \neq DE^2 + CE^2 = 25 + 13 = 38,$$
so $\triangle CDE$ is not a right triangle and hence $\angle CED$ is not a right angle.

Example 5.11

Suppose a square has diagonal of length d. Show that the square has area $\frac{d^2}{2}$. ♣

Solution

Consider the diagram below

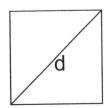

This diagonal divides the square into two congruent 45-45-90 triangles. (They are congruent by SAS.) Hence, using the ratio of sides in a 45-45-90 triangle, the square has side length
$$d \times \frac{1}{\sqrt{2}} = \frac{d\sqrt{2}}{2}.$$

Thus the area of the square is

$$\left(\frac{d\sqrt{2}}{2}\right)^2 = \frac{2d}{4} = \frac{d}{2}.$$

Example 5.12

(2007 AMC 8) In trapezoid $ABCD$, AD is perpendicular to DC, $AD = AB = 3$, and $DC = 6$. In addition, E is on DC, and BE is parallel to AD.

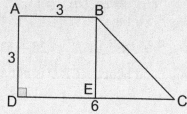

Find the area of $\triangle BEC$.

Solution

Since BE is parallel to AD we have that $ABED$ is a square with side length 3, so $DE = 3$. Therefore,

$$EC = DC - DE = 6 - 3 = 3.$$

Since $AD = BE = 3$ as well, the area of $\triangle BEC$ is

$$\frac{1}{2} \times 3 \times 3 = 4.5.$$

Example 5.13

The sides of an equilateral triangle are s. How long is an altitude of this triangle? What is the area of the triangle?

5.1 Example Questions

Solution

Note that drawing the altitude divides the triangle into two congruent 30-60-90 triangles with side lengths
$$\frac{s}{2}, \frac{s\sqrt{3}}{2}, s.$$
(Compare this to the diagram for showing a 30-60-90 triangle has sides in ratio $1 : \sqrt{3} : 2$ from above.) The altitude is the side length with length
$$\frac{s\sqrt{3}}{2}$$
and therefore the area of the triangle is
$$\frac{1}{2} \times \frac{s\sqrt{3}}{2} \times s = \frac{s^2\sqrt{3}}{4}.$$

Example 5.14

Draw the largest possible square inside an equilateral triangle, with one side of the square aligned with one side of the triangle. If the square has side length 6, find the side length of the equilateral triangle.

Solution

Note that the square and equilateral triangle form two 30-60-90 triangles as in the diagram below

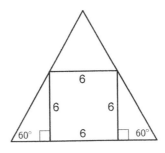

Note the bottom side of each 30-60-90 triangle is

$$6 \times \frac{1}{\sqrt{3}} = \frac{6}{\sqrt{3}} = 2\sqrt{3},$$

so the side length of the equilateral triangle is

$$6 + 2 \times 2\sqrt{3} = 6 + 4\sqrt{3}.$$

Example 5.15

(2005 AMC 8) In quadrilateral $ABCD$, sides $\overline{AB}$ and $\overline{BC}$ both have length 10, sides $\overline{CD}$ and $\overline{DA}$ both have length 17, and the measure of angle ADC is $60°$.

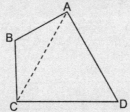

What is the length of diagonal $\overline{AC}$?

Solution

Note that an isosceles triangle with angle $60°$ lying between the equal sides implies that the triangle is equilateral. Therefore, $AC = DA = 17$.

5.2 Quick Response Questions

Problem 5.1 If one of the leg lengths of a right triangle is 20 and the other leg length is 21, what is the length of the hypotenuse?

Problem 5.2 If one of the leg lengths of a 45-45-90 triangle is 3, what is the length of the hypotenuse?

Problem 5.3 If the shorter leg length of a 30-60-90 triangle is 3, what is the length of the hypotenuse?

Problem 5.4 Consider the diagram below.

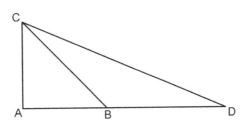

Suppose $AB = AC = 1$ and $\angle BAC = 90°$ and $\angle ADC = 22.5°$. Find BD.

Problem 5.5 Let *ABC* be a right triangle with leg lengths 8 and 15. All lengths of *ABC* are doubled to form new triangle *DEF*. Find the ratio of the area of *ABC* to the area of *DEF*.

5.3 Practice

Problem 5.6 Given two squares, show how to cut them up into pieces that can be combined to form one larger square. Your method should work no matter which two squares you are given. It is probably easiest to start with the squares side by side as in the diagram below.

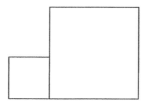

Problem 5.7 For each of the following, state whether it is possible to have a right triangle with the given side lengths. If it is possible, we call (a,b,c) a *Pythagorean Triple*.

(a) $6, 18, 21$

(b) $7, 24, 25$

(c) $8, 15, 17$

(d) 9, 18, 27

(e) 10, 24, 26

Problem 5.8 Let $ABCD$ be a rectangle with $AB = 18, BC = 5$. Let E be the point a third of the way from A to B on $\overline{AB}$. Is $\angle CED$ a right angle?

Problem 5.9 (1999 AMC 8) In trapezoid $ABCD$, the sides AB and CD are equal.

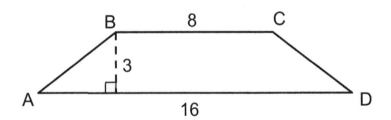

What is the perimeter of $ABCD$?

5.3 Practice

Problem 5.10 Let's work with isosceles right triangles.

(a) The legs of an isosceles right triangle are each 4 cm long. How long is an altitude drawn to the hypotenuse of this triangle?

(b) The altitude of an isosceles right triangle that meets the hypotenuse of the triangle has length 3 cm. What is the length of the shortest side of the triangle?

(c) If the area of an isosceles right triangle is 2 square inches, how long is the shortest side?

Problem 5.11 In the diagram, $\triangle ABC, \triangle DEF$ are two congruent isosceles right triangles. Given that $ADFC$ is a 4×2 rectangle, find the area of the shaded region.

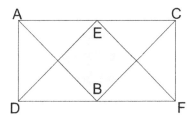

Problem 5.12 Mark P inside square $ABCD$, so that triangle ABP is equilateral. Let Q be the intersection of BP with diagonal AC. Triangle CPQ looks isosceles. Is this actually true?

Problem 5.13 Given a parallelogram $ABCD$, let P and Q be the points outside the parallelogram so that triangles CDP and BCQ are equilateral. Is the triangle APQ is equilateral?

Problem 5.14 (1999 AMC 8) Square $ABCD$ has sides of length 3. Segments CM and CN divide the square's area into three equal parts.

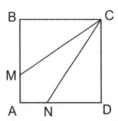

How long is segment CM?

5.3 Practice

Problem 5.15 Let ABC be a triangle given below:

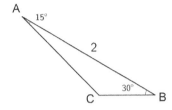

Determine the area of triangle ABC.

Problem 5.16 (2005 AMC 8) What is the perimeter of trapezoid $ABCD$?

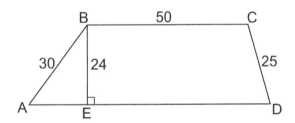

Problem 5.17 Draw the largest possible square inside an equilateral triangle, with one side of the square aligned with one side of the triangle. If the equilateral triangle has side length 6, find the side length of the square.

Problem 5.18 (2004 AMC 8) In the figure, $ABCD$ is a rectangle and $EFGH$ is a parallelogram.

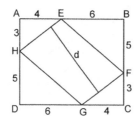

Using the measurements given in the figure, what is the length d of the segment that is perpendicular to $\overline{HE}$ and $\overline{FG}$?

Problem 5.19 The ratio of the hypotenuse of a right triangle to one of the other sides is $41 : 9$. What is the smallest possible area of the triangle if the length of each side of this triangle is an integer?

Problem 5.20 A triangle has sides measuring 37 cm, 37 cm and 24 cm. A second triangle not congruent to the first, is drawn with sides 37 cm, 37 cm and x cm, where x is a whole number. If the two triangles have equal areas, what is the value of x?

Copyright © ARETEEM INSTITUTE. All rights reserved.

5.3 Practice

Problem 5.21 (ZIML 2016) In $\triangle ABC$, $\angle B = 90°$, $AB = 8$, and $BC = 6$. Extend the line segment $\overline{BC}$ to point D so that the obtuse triangle $\triangle ACD$ is formed. If $CD = 9$, find the perimeter of $\triangle ACD$.

Problem 5.22 Let $A = (0,0), B = (1,1)$, and $C = (2,0)$ be coordinates. Find the angle measure of BCA?

Problem 5.23 (2015 AMC 8) In $\triangle ABC$, $AB = BC = 29$, and $AC = 42$. What is the area of $\triangle ABC$?

Problem 5.24 (2012 AMC 8) A square with area 4 is inscribed in a square with area 5, with one vertex of the smaller square on each side of the larger square. A vertex of the smaller square divides a side of the larger square into two segments, one of length a, and the other of length b.

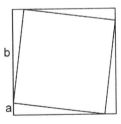

What is the value of ab?

Problem 5.25 (2015 AMC 8) One-inch squares are cut from the corners of this 5 inch square.

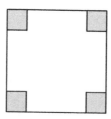

What is the area in square inches of the largest square that can be fitted into the remaining space?

6. Angles Are Special Too

In this chapter, we explore more about the special angles themselves.

The concepts introduced in this chapter directly correspond to Common Core Math Standards as shown in the following table.

6th Grade	6.RP.1, 6.RP.3, 6.EE.1, 6.EE.6, 6.G.1
7th Grade	7.EE.4, 7.G.1, 7.G.5, 7.G.6
8th Grade	8.EE.2, 8.G.6, 8.G.7, 8.G.8

In addition to the standards above, problems and concepts in this section will help strengthen understanding of the following domains.

6th Grade	6.RP, 6.EE, 6.G
7th Grade	7.RP, 7.EE, 7.G
8th Grade	8.EE, 8.G

Copyright © ARETEEM INSTITUTE. All rights reserved.

Special Angles

We call angles that are multiples of $30°$ ($30°, 60°, 90°, 120°, \ldots$) or multiples of $45°$ ($45°, 90°, 135°, \ldots$) special angles. Often the special triangles (30-60-90 or 45-45-90) are very helpful when dealing with special angles.

Example 6.1

Suppose we want to find the third side of the triangle below containing a $135°$ angle.

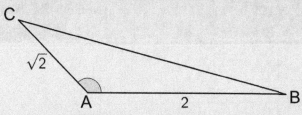

Note that $135° = 180° - 45°$, so we can extend BA to form a 45-45-90 triangle as shown below.

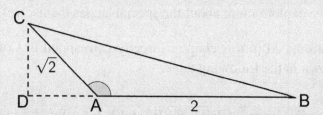

Since $\triangle ADC$ is a 45-45-90 triangle, we know that $AD = CD = 1$. Then using the Pythagorean theorem on right triangle BCD we have

$$BC^2 = 1^2 + (2+1)^2 = 10$$

so $BC = \sqrt{10}$.

Polygon

A shape made up of line segments is called a polygon. These line segments are called the sides of the polygon. For example, a triangle is a polygon with three sides.

Example 6.2 Equal Sides or Angles

- If all the sides of a polygon are equal we say it is an equilateral polygon.
- If all the angles in a polygon are equal we say it is an equiangular polygon.
- If a polygon is equilateral and equiangular we call it a regular polygon.

Example 6.3 Equilateral Triangle

An equilateral triangle has three equal sides and three equal angles, so it is a regular polygon.

Remark

We saw that an equilateral triangle must also be equiangular and an equiangular triangle must also be equilateral. This is not true for polygons with more than three sides, as we will see in the questions below.

Example 6.4 Introductory Areas

Given a polygon ABC or $ABCD$ we will use square brackets to denote area. Thus $[ABC]$ denotes the area of triangle ABC. Recall that:

- If a triangle ABC has base b and height h, we have $[ABC] = \frac{1}{2} \times b \times h$.
- If a rectangle $ABCD$ has base b and height h, we have $[ABCD] = b \times h$.

Copyright © ARETEEM INSTITUTE. All rights reserved.

> **Remark**
>
> To solve problems with polygons an accurate diagram always helps! If available, use a compass, ruler, or protractor to help make diagrams if you are stuck.
>
> Further, try to break a problem into smaller parts, hopefully involving special angle or the special 30-60-90 or 45-45-90 triangles.

6.1 Example Questions

> **Example 6.5**
>
> When working with polygons and regular polygons it is useful to have a chart summarizing the basic information. Your chart should include
> - Number of sides
> - Polygon name
> - Sum of interior angles
> - Sum of exterior angles
> - Each angle if the polygon is regular
> - Anything else you thing is useful
>
> Keep your chart for your own reference!

Solution

An example of one such chart is given below:

6.1 Example Questions

# of Sides	Polygon Name	Sum of Interior Angles	Sum of Exterior Angles	Each Angle (if Regular)
3	Triangle	180°	360°	60°
4	Quadrilateral	360°	360°	90°
5	Pentagon	540°	360°	108°
6	Hexagon	720°	360°	120°
7	Heptagon	900°	360°	$\left(\frac{900}{7}\right)°$
8	Octagon	1080°	360°	135°
10	Decagon	1440°	360°	144°
12	Dodecagon	1800°	360°	150°

Note that the sum of interior angles increases by 180° for every side added. Therefore in general a polygon with n sides has a sum of interior angles of $180(n-2)$ degrees.

Example 6.6

Use 6 equilateral triangles to form a hexagon $ABCDEF$. Show (with justification) hexagon $ABCDEF$ is regular. Further, calculate the angle AED.

Solution

It is clear that each side has the same length. Each angle in the hexagon is made up of two 60° angles (from the equilateral triangle), hence each angle is 120°. Thus the hexagon is equilateral and equiangular and hence regular.

For $\angle AED$, first note that since the hexagon is regular we know $\triangle AFE$ is isosceles with $\angle AFE = 120°$. Thus,
$$\angle FEA = \frac{1}{2} \times (180 - 120) = 30°$$
and
$$\angle AED = \angle FED - \angle FEA = 120 - 30 = 90°.$$

Example 6.7

Three non-overlapping regular plane polygons all have sides of length 1. The polygons meet at a point A in such a way that the sum of the three interior angles at A is $360°$. Among the three polygons, one is a triangle and one is a dodecagon. Find the remaining polygon.

Solution

The remaining angle is
$$360 - 60 - 150 = 150°,$$
which belongs to an another dodecagon (12 sides).

Example 6.8

Let $ABCDEF$ be a regular hexagon, and let $EFGHI$ be a regular pentagon. Find all possible values of measure of $\angle GAF$.

Solution

Note that the pentagon can either be in the hexagon or outside the hexagon as in the diagram below with the relevant points labeled.

6.1 Example Questions

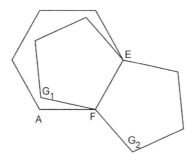

Note both $\triangle AFG_1$ and $\triangle AFG_2$ are isosceles. Recall that interior angles of a regular hexagon are $120°$ and those of a regular pentagon are $108°$. If the pentagon is inside the hexagon, then
$$\angle AFG_1 = 120° - 108° = 12°,$$
so
$$\angle G_1AF = \angle AG_1F = (180° - 12°) \div 2 = 84°.$$
If the pentagon is outside the hexagon, we similarly have that
$$\angle AFG_2 = 360° - 120° - 108° = 132°,$$
so
$$\angle G_2AF = \angle AG_2F = (180° - 132°) \div 2 = 24°.$$

Example 6.9

Suppose five squares are attached to a regular pentagon as in the diagram below.

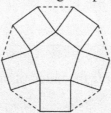

Connecting vertices of the squares as shown forms a decagon. Is this decagon equiangular? Is it equilateral?

Solution

First note since we are attaching squares to a regular pentagon, all the marked segments in the diagram below are equal.

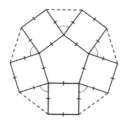

Therefore all the outside triangles are isosceles. Further, all the marked angles are equal, with measure
$$360° - 90° - 108° - 90° = 72°.$$
Hence these triangles have one angle of $72°$ and the other two angles are
$$(180° - 72°) \div 2 = 54°.$$
Hence all the interior angles of the decagon are
$$90° + 54° = 144°$$
so the decagon is equiangular. However, as the triangles have angles $72°, 54°, 54°$ they are not equilateral triangles, so the decagon is not equilateral.

Example 6.10

Explain where and how to cut off the sides of an equilateral triangle to form a regular hexagon.

Solution

If we place points dividing each side into thirds and draw parallel line segments we get the diagram below.

Copyright © ARETEEM INSTITUTE. All rights reserved.

6.1 Example Questions

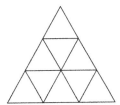

All of these triangles are congruent equilateral triangles and it is easy to see that removing the outer three results in a regular hexagon.

Example 6.11

Find the area of a regular hexagon with side length 12.

Solution

The hexagon is made up of 6 equilateral triangles, each with an area of

$$\frac{1}{2} \times 12 \times 6\sqrt{3} = 36\sqrt{3}.$$

Hence the area is

$$6 \times 36\sqrt{3} = 216\sqrt{3}.$$

Example 6.12

Mark Y inside regular pentagon $PQRST$, so that PQY is equilateral. Is RYT straight?

Solution

Using interior angles for a regular pentagon and an equilateral triangle first calculate

$$\angle YPT = \angle YQR = 108° - 60° = 48°.$$

Hence as $\triangle YPT$ and $\triangle YQR$ are isosceles,

$$\angle PYT = \angle RYQ = (180° - 48°) \div 2 = 66°.$$

so

$$\angle RYT = 66° + 60° + 66° = 168° \neq 180°$$

so RYT is not a straight line.

Example 6.13

Attach a regular pentagon $ABDEF$ to the side of an equilateral triangle ABC. Calculate $\angle CDE$.

Solution

Note $\triangle BCD$ is an isosceles triangle, and

$$\angle CBD = 60 + 108 = 168°.$$

Hence

$$\angle BDC = (180 - 168) \div 2 = 6°,$$

so

$$\angle CDE = 108 - 6 = 102°.$$

6.1 Example Questions

Example 6.14

(2007 AMC 8) A unit hexagram is composed of a regular hexagon of side length 1 and its 6 equilateral triangular extensions, as shown in the diagram.

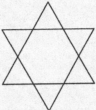

What is the ratio of the area of the extensions to the area of the original hexagon?

Solution

Recall a regular hexagon can be decomposed into six equilateral triangles. Thus, each of the equilateral triangles on the outside of the hexagon can be "folded in" to perfectly cover the hexagon. Hence the ratio is $1 : 1$.

6.2 Quick Response Questions

Problem 6.1 Give an example of an equiangular polygon that is not equilateral.

Problem 6.2 Give an example of an equilateral polygon that is not equiangular.

Problem 6.3 Inside regular pentagon $ABCDE$ is marked point F so that triangle ABF is equilateral. Decide whether or not quadrilateral $ABCF$ is a parallelogram, and give your reasons.

Problem 6.4 A regular nonagon has nine sides. Find the interior angle sum, the exterior angle sum, and the interior angle measure of a nonagon.

Problem 6.5 A regular icosagon has twenty sides. Find the interior angle sum, the exterior angle sum, and the interior angle measure of a nonagon.

Copyright © ARETEEM INSTITUTE. All rights reserved.

6.3 Practice

Problem 6.6 Use 6 equilateral triangles to form a hexagon $ABCDEF$, then find the area of triangle AED in terms of the total area.

Problem 6.7 Equilateral triangles BCP and CDQ are attached to the outside of regular pentagon $ABCDE$. Is quadrilateral $BPQD$ a parallelogram? Justify your answer.

Problem 6.8 Three non-overlapping regular plane polygons all have sides of length 1. The polygons meet at a point A in such a way that the sum of the three interior angles at A is $360°$. Thus the three polygons form a new polygon P (not necessarily convex) with A as an interior point. Suppose two of the polygons are pentagons. Find the perimeter of P.

Problem 6.9 Let $ABCDEFGH$ be a regular octagon, and let $GHIJKL$ be a regular hexagon. Find all possible values of measure of $\angle IAH$.

Copyright © ARETEEM INSTITUTE. All rights reserved.

Problem 6.10 Suppose that *DRONE* is a regular pentagon, and that *DRU, ROC, ONL, NEA*, and *EDI* are equilateral triangles attached to the outside of the pentagon. Is *IUCLA* a regular pentagon?

Problem 6.11 The equiangular convex hexagon *ABCDEF* has $AB = 10$, $BC = 10$, $CD = 8$, and $DE = 8$. Find $[ABCDEF]$.

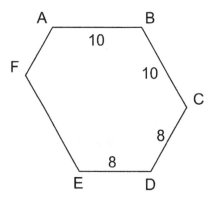

Problem 6.12 Find the area of a regular octagon with side length 12.

Problem 6.13 Mark *Y* inside regular hexagon *PQRSTU*, so that *PQY* is equilateral. Is *RYU* straight?

6.3 Practice

Problem 6.14 Suppose that *EARTH* is a regular pentagon and regular pentagons *ANGLE* and *RAPID* are attached to the outside of the pentagon. Show that N, G, P, I are lie on a single line.

Problem 6.15 A stop sign — a regular *octagon* — can be formed from a square sheet of metal by making four straight cuts that snip off the corners. If we want an octagon with sides of length $\sqrt{2}$, how large does the side of the original square need to be?

Problem 6.16 (2009 AMC 8) Construct a square on one side of an equilateral triangle. On one non-adjacent side of the square, construct a regular pentagon, as shown. On one non-adjacent side of the pentagon, construct a hexagon. Continue to construct regular polygons in the same way, until you construct an octagon.

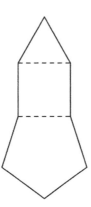

How many sides does the resulting polygon have?

Problem 6.17 A triangle has a 60° angle and a 45° angle, and the side opposite the 45° angle has length 12. How long is the side opposite the 60° angle?

Problem 6.18 Let $ABCD$ be a square with area 3 and let EF be a line segment that divides the square into two congruent trapezoids. What is the perimeter of trapezoid $EFBC$?

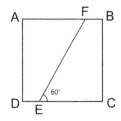

Problem 6.19 (2015 AMC 8) In the given figure hexagon $ABCDEF$ is equiangular, $ABJI$ and $FEHG$ are squares with areas 18 and 32 respectively, $\triangle JBK$ is equilateral and $FE = BC$.

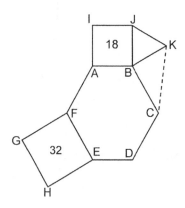

What is the area of $\triangle KBC$?

6.3 Practice

Problem 6.20 A stop sign — a regular *octagon* — can be formed from a square sheet of metal by making four straight cuts that snip off the corners. If we have a square with sides of length $\sqrt{2}$, what is the side length of the resulting octagon?

Problem 6.21 Let *ABC* be a triangle given below:

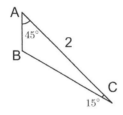

Find the perimeter of the triangle.

Problem 6.22 From the previous question, determine the area of triangle *ABC*.

Problem 6.23 Let *ABC* be an equilateral triangle that contains a 45-45-90 triangle *DEF* with hypotenuse parallel to a side of the equilateral triangle given below:

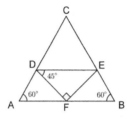

If $AB = 1$, find DF.

Problem 6.24 Using the above figure, find the area of triangle *DEF*.

Problem 6.25 Using the above figure, find the side ratio of triangle *DAF*.

7. Discovering Areas

So far we have seen some basic area formulas. It is a good idea to understand how the different formulas relate to each other. In fact, we will see in this chapter that most area formulas can be build up from just the formula for the area of a rectangle.

The concepts introduced in this chapter directly correspond to Common Core Math Standards as shown in the following table.

6th Grade	6.RP.1, 6.RP.3, 6.EE.1, 6.EE.6, 6.G.1, 6.G.3
7th Grade	7.EE.4, 7.G.1, 7.G.6
8th Grade	8.EE.2, 8.G.6, 8.G.7, 8.G.8

In addition to the standards above, problems and concepts in this section will help strengthen understanding of the following domains.

6th Grade	6.RP, 6.EE, 6.G
7th Grade	7.RP, 7.EE, 7.G
8th Grade	8.EE, 8.G

Copyright © ARETEEM INSTITUTE. All rights reserved.

Example 7.1 Area of a Parallelogram

A parallelogram has base b and height h has area $b \times h$.

To understand this formula, consider cutting a triangle at one end of the parallelogram and moving it to the other end of the parallelogram as in the diagram below.

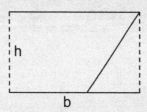

This results in a rectangle (still with base b and height h) with area $b \times h$.

Example 7.2 Area of a Triangle

The area of a triangle with base b and height h is $\frac{1}{2} \times b \times h$ because two copies of a triangle can be combined to form a parallelogram with base b and height h.

Remark

Note using these formulas we have the simple but useful fact: "Triangles (or parallelograms) with equal bases and equal heights have the same area."

7.1 Example Questions

Example 7.3

Using only the basics about parallel lines and congruent/similar triangles and the fact that the area of the rectangle is bh, prove the following: (Note: once a fact is proven below, you can use it in later parts.)
(a) The area of a parallelogram is bh.
(b) The area of a triangle is $\frac{1}{2}bh$.
(c) The area of a trapezoid is $\frac{b_1+b_2}{2}h$.

Solution

For the parallelogram formula, cutting a triangle at one end of the parallelogram and moving it to the other results in a $b \times h$ rectangle:

For the triangle formula, note that two copies of any triangle can be combined to form a parallelogram with base b and height h.

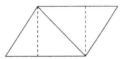

Lastly, for the trapezoid formula, note that a diagonal of a trapezoid breaks the trapezoid into two triangles, each with height h and having bases b_1 and b_2.

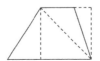

Example 7.4

(2013 AMC 8) Squares $ABCD$, $EFGH$, and $GHIJ$ are equal in area. Points C and D are the midpoints of sides IH and HE, respectively.

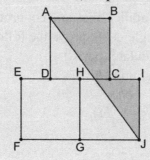

What is the ratio of the area of the shaded pentagon $AJICB$ to the sum of the areas of the three squares?

Solution

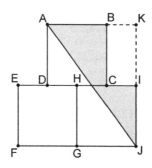

Without loss of generality, let $s = 2$ to be the side length of the square. Extend the length of AB and IJ so that they intersect at point K and create a new rectangular region $CIKB$.

The area of triangle AKJ is
$$\frac{3 \times 4}{2} = 6$$
and the area of rectangle $CIKB$ is
$$2 \times 1 = 2.$$

The area of the shaded pentagon is
$$6 - 2 = 4.$$

Copyright © ARETEEM INSTITUTE. All rights reserved.

7.1 Example Questions

The ratio of the shaded area to the combined area of the three squares is

$$\frac{4}{3 \times 2^2} = \frac{1}{3}.$$

Remark

In this book, we use $[ABC]$ to represent the area of $\triangle ABC$, and $[DEFGH]$ the area of pentagon $DEFGH$, etc. Using this notation, in the previous example, $[AKJ] = 6$, $[CIKB] = 2$, and the shaded region $[ABCIJ] = 4$.

Example 7.5

Explain why the following are true. Note these facts will be very useful for later problems!

(a) Triangles (or parallelograms) with equal bases and equal heights have the same area.

(b) In $\triangle ABC$, let D be the midpoint of side $\overline{BC}$ and connect $\overline{AD}$, as in the diagram below.

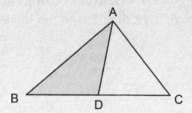

Prove that $[ABD] = [ACD]$ (that is, $\triangle ABD$ has the same area as $\triangle ACD$).

(c) Let $ABCD$ be a parallelogram and E be any point on side $\overline{CD}$ as shown.

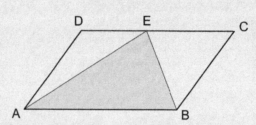

Prove that $[ABE] = \frac{1}{2}[ABCD]$.

(d) In $\triangle ABC$, let D be any point on side $\overline{BC}$ as shown below.

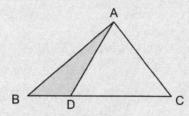

Prove that the ratios $\dfrac{[ABD]}{[ACD]} = \dfrac{BD}{CD}$.

7.1 Example Questions

Solution

For (a), this is clear from the formula for the area of a triangle.

For (b), note that these two triangles have the same height and the same base (as D is the midpoint).

For (c), regardless of where E is on $\overline{CD}$, $\triangle ABE$ has the same height and base as the parallelogram.

For (d), both triangles share the same height, say h. Then

$$[ABD] = \frac{1}{2} h \times AB, [ACD] = \frac{1}{2} h \times CD$$

and the result follows after canceling.

Example 7.6

Use three segments of length 4 cm, 6 cm, and 8 cm as the bases and altitude (not necessarily in the given order) to make a trapezoid. Trapezoids of three different possible areas can be made. Which one is the largest, and what is its area?

Solution

Calculate the areas of trapezoids using each of the 3 lengths as altitude and the remaining two as bases. The areas are:

If 4 cm is the height:
$$\frac{1}{2}(6+8) \times 4 = 28;$$

If 6 cm is the height:
$$\frac{1}{2}(4+8) \times 6 = 36;$$

If 8 cm is the height:
$$\frac{1}{2}(4+6) \times 8 = 40.$$

Thus the largest area is obtained by using the longest segment has the height.

Copyright © ARETEEM INSTITUTE. All rights reserved.

Example 7.7

The parallelogram $ABCD$ has area 48cm^2, $AE = 8\text{cm}$, $CE = 4\text{cm}$. Find the area of the shaded region.

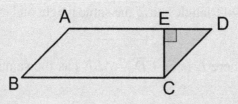

Solution

Note that
$$AD = 48 \div 4 = 12$$
and
$$ED = 12 - 8 = 4.$$
Thus, the area is determined as follows
$$[DEC] = \frac{1}{2} \times 4 \times 4 = 8\text{cm}^2.$$

Example 7.8

(2008 AMC 8) In the figure, the outer equilateral triangle has area 16, the inner equilateral triangle has area 1, and the three trapezoids are congruent.

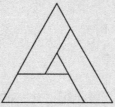

What is the area of one of the trapezoids?

7.1 Example Questions

Solution

Given that a figure of area 16 has three congruent trapezoids and one equilateral triangle with area 1, the combined area of the three congruent trapezoids is 15.

This implies that the area of one trapezoid is
$$15 \div 3 = 5.$$

Example 7.9

(2010 AMC 8) The diagram shows an octagon consisting of 10 unit squares.

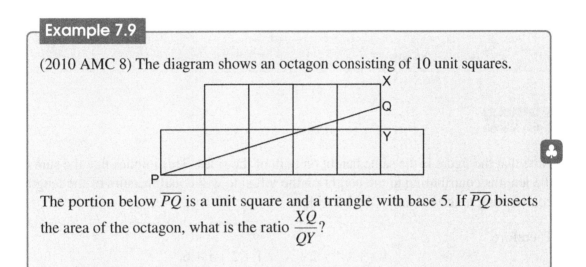

The portion below $\overline{PQ}$ is a unit square and a triangle with base 5. If $\overline{PQ}$ bisects the area of the octagon, what is the ratio $\dfrac{XQ}{QY}$?

Solution

Since the line bisects the area of the octagon, we observe that each piece has area 5.

The lower portion is a region consisting of a triangle and a unit square. Since the unit square has area 1, the triangle must have area 4.

Therefore, if we let h represent the height of the triangle, it follows that
$$\frac{1}{2} \times 5 \times h = 4$$
or equivalently, $h = \frac{8}{5}$. The length of QY is
$$\frac{8}{5} - 1 = \frac{3}{5}.$$

The length of XQ is $\frac{2}{5}$ and the ratio of the lengths of XQ and QY is $\frac{2}{5} : \frac{3}{5}$ or $\frac{2}{3}$.

Example 7.10

(2012 AMC 8) In the diagram, all angles are right angles and the lengths of the sides are given in centimeters. Note the diagram is not drawn to scale. What is X in centimeters?

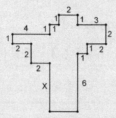

Solution

Note that the figure is the same height on both of the sides. This implies that the sum of the lengths contributing to the height on the left side will equal the sum of the lengths contributing to the height on the left side.

Therefore,
$$1+1+1+2+X = 1+2+1+6$$
implies that $X = 5$.

Example 7.11

A six-pointed star is formed in the following way: Six points A, B, C, D, E and F are equally spaced on a circle and triangles ACE and BDF are formed. The combined regions of $\triangle ACE$ and $\triangle BDF$ give the star. If $AC = 3$, what is the area of the star?

Solution

The points are equally spaced, so the star is made up of 6 equilateral triangles and one regular hexagon. A regular hexagon can be decomposed into another 6 equilateral triangles congruent to the other equilateral triangles as in the diagram below.

Copyright © ARETEEM INSTITUTE. All rights reserved.

7.1 Example Questions

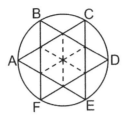

We know $AC = 3$, so the side length of each equilateral triangle is $3 \div 3 = 1$. Since there are 12 equilateral triangles in total, the area of the star is

$$12 \times \frac{1}{2} \times \frac{\sqrt{3}}{2} \times 1 = 3\sqrt{3}.$$

Example 7.12

(2014 AMC 8) Rectangle $ABCD$ and right triangle DCE have the same area. They are joined to form a trapezoid, as shown.

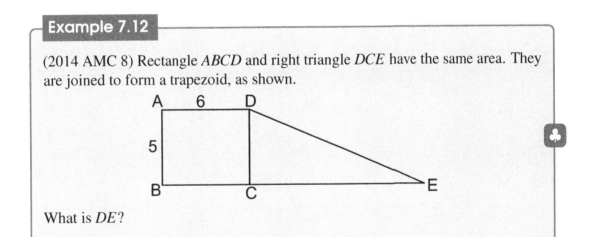

What is DE?

Solution

It is clear to see that the area of $ABCD$ is

$$AB \times AD = 5 \times 6 = 30.$$

Given that the area of the rectangle is equal to the area of the triangle, since $CD = AB = 5$ and the area of triangle CDE is 30,

$$\frac{1}{2} \times 5 \times CE = 30.$$

Hence we know that
$$CE = 30 \times \frac{2}{5} = 12.$$
By Pythagorean Theorem, the length of *DE* is
$$\sqrt{5^2 + 12^2} = \sqrt{169} = 13.$$

7.2 Quick Response Questions

Problem 7.1 Let $ABCD$ be a square and let $EFGH$ be a rectangle such that $[ABCD] = [EFGH]$. Suppose that the side length of the rectangle is twice the side length of the square. Find the ratio of the perimeter of the square to the perimeter of the rectangle.

Problem 7.2 What is the area of a regular hexagon with side length 3?

Problem 7.3 What is the area of a regular octagon with side length 3?

Problem 7.4 If the area of parallelogram $ABCD$ is 10 and the area of triangle CDE is 2, what is the area of trapezoid $ABCE$?

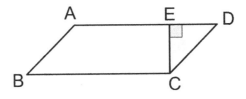

Copyright © ARETEEM INSTITUTE. All rights reserved.

Problem 7.5 If the area of the rectangle is 40, the area of the smaller white triangle is 5, and the area of the big white triangle is 20, what is the area of the shaded region?

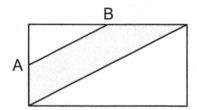

7.3 Practice

Problem 7.6 Suppose you make a parallelogram with base 5cm and height 4cm.

(a) Find the area of the parallelogram.

(b) Show it is possible to have different perimeters for a parallelogram as above. Is there a largest perimeter that is possible? What about a smallest?

Problem 7.7 The parallelogram $ABCD$ has area 300cm^2 and E is the midpoint of $\overline{AD}$. Find the area of the shaded region.

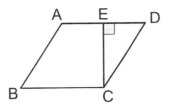

Problem 7.8 A garden of rectangular shape is shown in the diagram. The shaded regions are grass, and the unshaded regions are empty spaces in the shape of six congruent rhombi. Find the ratio between the areas of the grass and empty regions.

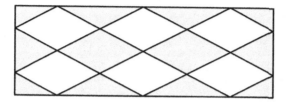

Problem 7.9 In the diagram, points A, B, C, D are the midpoints of their respective sides. Compute the ratio of the area of the shaded region and the whole rectangle.

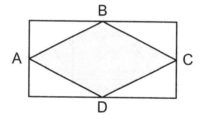

Problem 7.10 In the parallelogram $ABDC$, points E and F are the midpoints of sides $\overline{AD}$ and $\overline{DC}$ respectively. Which triangles have the same area as $\triangle BFC$?

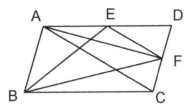

Problem 7.11 In the rectangle $ABCD$, the area of $\triangle AOB$ is 6 cm². Let O be a point inside rectangle such that the area of $\triangle DOC$ is $1/3$ of the area of the rectangle. Find the area of rectangle $ABCD$.

Problem 7.12 (2015 AMC 8) Point O is the center of the regular octagon $ABCDEFGH$, and X is the midpoint of the side $\overline{AB}$.

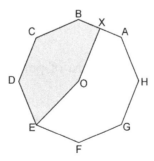

What fraction of the area of the octagon is shaded?

Problem 7.13 Let *ABCD* be a parallelogram, as in the diagram. Compare the shaded regions △*ABF* and △*DEF*, which one has the larger area?

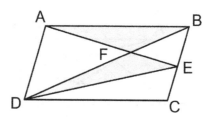

Problem 7.14 In the diagram, points *A* and *B* are the midpoints of their respective sides. Compute the ratio of the area of the shaded region and the whole rectangle.

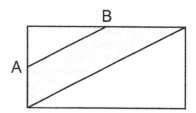

Problem 7.15 (2005 AMC 8) The area of polygon *ABCDEF* is 52 with $AB = 8$, $BC = 9$ and $FA = 5$. What is $DE + EF$?

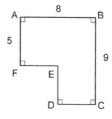

Copyright © ARETEEM INSTITUTE. All rights reserved.

7.3 Practice

Problem 7.16 A garden of rectangular shape is shown in the diagram. The shaded regions are grass, and the unshaded regions are empty spaces in the shape of four congruent hexagons. Find the ratio between the areas of the grass and empty regions.

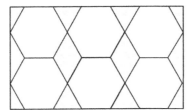

Problem 7.17 In the diagram, $\triangle ABC, \triangle DEF$ are two congruent isosceles right triangles. Given that $AB = 9, EC = 3$, find the area of the shaded region.

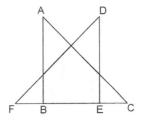

Problem 7.18 (2000 AMC 8) Triangles ABC, ADE, and EFG are all equilateral. Points D and G are midpoints of $\overline{AC}$ and $\overline{AE}$, respectively. If $AB = 4$, what is the perimeter of figure $ABCDEFG$?

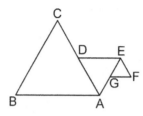

Problem 7.19 In the given trapezoid $ABCD$, there are 8 triangles. Among them, the pair $\triangle ABC$ and $\triangle DBC$ have the same area. How many other pairs have the same areas? List the pairs with the same areas.

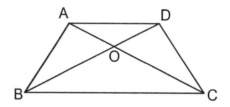

7.3 Practice

Problem 7.20 (2011 AMC 8) Quadrilateral $ABCD$ is a trapezoid, $AD = 15$, $AB = 50$, $BC = 20$, and the altitude is 12.

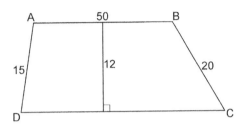

What is the area of the trapezoid?

Problem 7.21 Let $ABCD$ be a parallelogram with $[ABCD]$. Let P be a point in the interior of $ABCD$. Show that $[ABP] + [CDP] = [ABCD]/2$.

Problem 7.22 Let $ABCD$ be a parallelogram, as in the diagram. Suppose E is the midpoint of $\overline{BC}$. Find the area of the shaded regions $\triangle ABF$ and $\triangle DEF$ in terms of the entire area of the parallelogram.

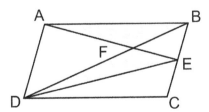

Copyright © ARETEEM INSTITUTE. All rights reserved.

Problem 7.23 If regular hexagon *ABCDEF* with side length 2 can be reinterpreted as 6 equilateral triangles, what is the area of *ABCDEF*?

Problem 7.24 (2006 AMC 8) The letter T is formed by placing two 2×4 inch rectangles next to each other, as shown. What is the perimeter of the T, in inches?

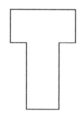

Problem 7.25 Find the perimeter of the quadrilateral formed by the coordinates $A = (0,0)$, $B = (4,3)$, $C = (6,3)$, and $D = (10,0)$?

8. Conquering Areas

Combining all the facts and formulas from the previous chapter allows us to solve many different problems involving areas!

The concepts introduced in this chapter directly correspond to Common Core Math Standards as shown in the following table.

6th Grade	6.RP.1, 6.RP.3, 6.EE.1, 6.EE.6, 6.G.1, 6.G.3
7th Grade	7.EE.4, 7.G.1, 7.G.6
8th Grade	8.EE.2, 8.G.6, 8.G.7, 8.G.8

In addition to the standards above, problems and concepts in this section will help strengthen understanding of the following domains.

6th Grade	6.RP, 6.EE, 6.G
7th Grade	7.RP, 7.EE, 7.G
8th Grade	8.EE, 8.G

Copyright © ARETEEM INSTITUTE. All rights reserved.

8.1 Example Questions

Example 8.1

All the rectangles in the following diagrams are squares. The lengths of the segments are marked. Find the area of the shaded regions in each diagram.

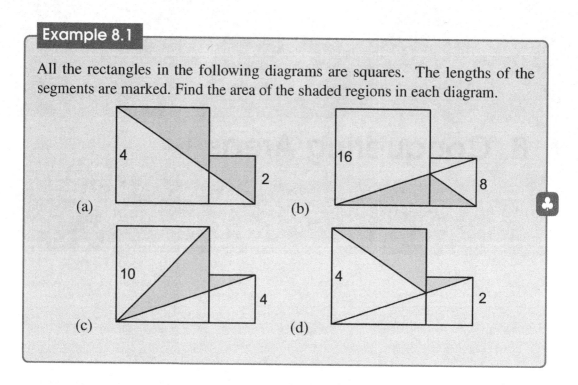

Solution

For (a), the shaded region is the two squares minus the unshaded triangle with base 6 and height 4, so the area is

$$4^2 + 2^2 - \frac{1}{2} \times 4 \times 6 = 8.$$

For (b), the shaded region is a triangle with base 24 and height 8 minus a triangle with base 8 and height 8, so the area is

$$\frac{1}{2} \times 24 \times 8 - \frac{1}{2} \times 8 \times 8 = 64.$$

For (c), the shaded region is the two squares minus triangles with base 10 and height 10 and base 14 and height 4, so the area is

$$10^2 + 4^2 - \frac{1}{2} \times 10 \times 10 - \frac{1}{2} \times 14 \times 4 = 38.$$

Copyright © ARETEEM INSTITUTE. All rights reserved.

8.1 Example Questions

For (d), the shaded region is the two squares minus triangles with base 4 and height 4 and base 6 and height 2, so the area is

$$4^2 + 2^2 - \frac{1}{2} \times 4 \times 4 - \frac{1}{2} \times 6 \times 2 = 6.$$

Example 8.2

(2004 AMC 8) What is the area enclosed by the geoboard quadrilateral below?

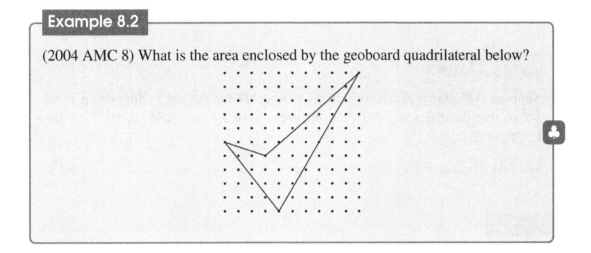

Solution

Consider the 10×10 grid divided into 5 regions as in the diagram below.

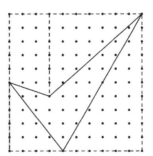

Hence the area we want can be computed by taking the area of the entire square board and subtracting from it the area of a triangle with base and height measurements 4 and 5 respectively, the area of a triangle with base and height measurements 6 and 10 respectively, the area of a triangle with base and height measurements 6 and 7 respectively, and the area of a trapezoid with base measurements 5 and 6 and height measurement 3.

Therefore, the desired area is

$$10^2 - \left(\frac{1}{2} \times 4 \times 5 + \frac{1}{2} \times 6 \times 10 + \frac{1}{2} \times 6 \times 7 + \frac{1}{2} \times (5+6) \times 3\right) = \frac{45}{2}.$$

Example 8.3

Suppose $ABCD$ is a parallelogram with base AD and height 5. Suppose E is on $\overline{BC}$ so that the difference between the areas $[ADCE]$ and $[ABE]$ is 15. Find the length of EC.

Solution

Draw $\overline{EF}$ parallel to $\overline{CD}$ with F on $\overline{AD}$ as drawn below.

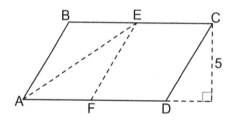

Note that $ABEF$ is another parallelogram, so $\triangle ABE$ and $\triangle EFA$ are congruent and have the same area. Therefore,

$$[ECDF] = [ECDA] - [EFA] = [ECDA] - [ABE] = 15.$$

Since $[ECDF] = 5 \times EC$ we have

$$EC = 15 \div 5 = 3.$$

Copyright © ARETEEM INSTITUTE. All rights reserved.

8.1 Example Questions

Example 8.4

In the diagram below, there are 36 rectangular grid points, evenly spaced, and the distance between each pair of adjacent points is 1. Find the area of $\triangle ABC$.

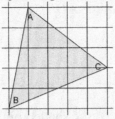

Solution

The square surrounding $\triangle ABC$ is 5×5 so has area 25. We subtract off the areas of the three triangles outside ABC:

$$25 - \frac{1}{2} \times 1 \times 5 - \frac{1}{2} \times 5 \times 2 - \frac{1}{2} \times 3 \times 4 = \frac{23}{2}.$$

Example 8.5

In the diagram below, there are 21 grid points arranged in equilateral triangles, equally spaced. The *area* of each small equilateral triangle formed by 3 adjacent grid points is 1. Find the area of $\triangle ABC$.

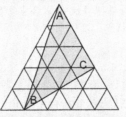

Solution

Call the entire triangle ADE (with C on $\overline{AE}$). $\triangle ABC$ and $\triangle ABE$ have the same height, so

$$[ABC] : [ABE] = 3 : 5 = 12 : 20.$$

Copyright © ARETEEM INSTITUTE. All rights reserved.

Similarly,
$$[ABE] : [ADE] = 4 : 5 = 20 : 25.$$
Combining these two we have
$$[ABC] : [ADE] = 12 : 25.$$
As $[ADE] = 25$, we have $[ABC] = 12$.

Example 8.6

(2007 AMC 8) What is the area of the shaded pinwheel shown in the 5×5 grid?

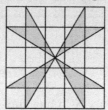

Solution

Firstly, note that the area of the square around the pinwheel is 25. The area of the pinwheel is equal to the area of the square minus the area of 4 unit squares (corners) minus the area of 4 triangles with height 2.5 and base 3. Therefore, the area of the pinwheel is
$$25 - \left(4 + 4 \times \frac{1}{2} \times 3 \times 2.5\right) = 6.$$

Example 8.7

In the diagram, $\triangle ABC$ and $\triangle DEF$ are two congruent isosceles right triangles. Given that $AB = 6, EC = 2$, find the area of the shaded region.

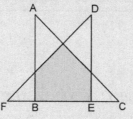

Copyright © ARETEEM INSTITUTE. All rights reserved.

8.1 Example Questions

Solution

Label the resulting pentagon $BEGHI$. We have $AB = BC = 6$, so

$$BE = 6 - 2 = 4.$$

Therefore, $FB = 2$ so $CF = 8$. Now note $\triangle CFH$ is a 45-45-90 triangle with hypotenuse 8. Hence its sides are length

$$8 \times \frac{1}{\sqrt{2}} = \frac{8}{\sqrt{2}} = 4\sqrt{2},$$

so it has area

$$[CFH] = \frac{1}{2} \times 4\sqrt{2} \times 4\sqrt{2} = 16.$$

Similarly,

$$[CEG] = [FBI] = \frac{1}{2} \times 2 \times 2 = 2.$$

Hence, the shaded pentagon has area

$$16 - 2 \times 2 = 12.$$

Example 8.8

Suppose $\triangle ABC$ with E on $\overline{AB}$ and D on $\overline{AC}$ such that $AE = \frac{1}{3}AB, AD = \frac{1}{2}AC$. If $[AED] = 2$, find the area of $\triangle ABC$.

Solution

We have $[ABD] = 3[AED]$ (same height) and similarly $[ABC] = 2[ABD]$, so

$$[ABC] = 6[AED] = 2 \times 6 = 12.$$

Example 8.9

Parallelogram *ABCD* is shown below, where triangle *BCE* is a right isosceles triangle and *A* is the midpoint of $\overline{BE}$. Given that $[ABCF] - [DFC] = 4$, find the area of *ABCD*.

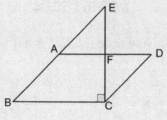

Solution

Since $\triangle BCE$ and $\triangle AFE$ are 45-45-90 triangles and *A* is a midpoint, we get that *F* is the midpoint of $\overline{AD}$. From here we get that

$$[ABCF] = \frac{3}{4} \times [ABCD], [DFC] = \frac{1}{4} \times [ABCD],$$

so

$$[ABCF] - [DFC] = \frac{1}{2} \times [ABCD].$$

Hence $[ABCD] = 8$.

Example 8.10

(1999 AMC 8) Points B, D, and J are midpoints of the sides of right triangle ACG. Points K, E, I are midpoints of the sides of triangle JDG, etc.

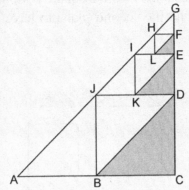

If the dividing and shading process is done 100 times (the first three are shown) and $AC = CG = 6$, what is the total area of the shaded triangles rounded to the nearest integer?

Solution

Notice that in every iteration, we construct three triangles with one of them shaded. If we repeat this iteration many times, we observe that the area shaded will be approximately $\frac{1}{3}$ of the total area of the triangle. Therefore,

$$\frac{1}{3}\left(\frac{1}{2} \times 6 \times 6\right) = 6$$

square units will be shaded when the process goes on indefinitely.

8.2 Quick Response Questions

Problem 8.1 Two people are building (tiny) houses. Both people will have triangular roofs that are 5m long (for each half), with a square bottom of the house. The first plans to have a house 8m wide, and the second plans to have a house 6m wide, as shown below:

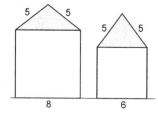

How tall is each house?

Problem 8.2 Using the above diagram, find the amount of attic space in each house. (That is the shaded region.)

Problem 8.3 Find the total amount of space in each house (including the attic).

Problem 8.4 Suppose we have a right triangle such that $AC = 3$, $BC = 4$, and $\angle C = 90°$. What is the area of this triangle?

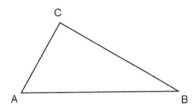

Problem 8.5 Now suppose we have the same right triangle such that $AC = 3$, $BC = 4$, and $\angle C = 90°$. What is the length of altitude CD?

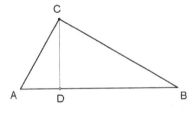

8.3 Practice

Problem 8.6 All the rectangles in the following diagrams are squares, except for (d), where the triangles are isosceles right triangles. The lengths of the segments are marked. Find the area of the shaded regions in each diagram.

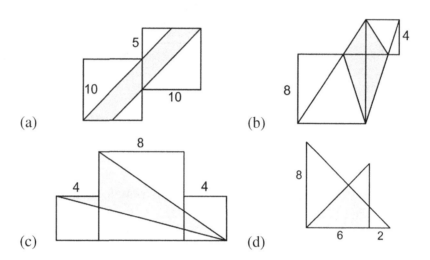

Problem 8.7 In parallelogram $ABCD$, M and N are midpoints of sides $\overline{AB}$ and $\overline{BC}$ respectively. Given that $[DMN] = 9$, find the area of $ABCD$.

Problem 8.8 In the diagram below, there are 36 rectangular grid points, evenly spaced, and the distance between each pair of adjacent points is 1. Find the area of quadrilateral *ABCD*.

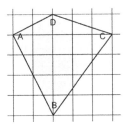

Problem 8.9 In the diagram below, there are 21 grid points arranged in equilateral triangles, equally spaced. The *area* of each small equilateral triangle formed by 3 adjacent grid points is 1. Find the area of quadrilateral *ABCD*.

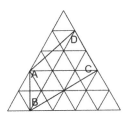

Problem 8.10 (2015 AMC 8) A triangle with vertices as $A = (1,3)$, $B = (5,1)$, and $C = (4,4)$ is plotted on a 6×5 grid.

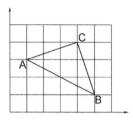

What fraction of the grid is covered by the triangle?

Problem 8.11 Suppose $\triangle ABC$ with D on $\overline{AB}$, G on $\overline{AC}$, and E, F on $\overline{BC}$. Suppose further $AD = \frac{1}{2}AB, BE = EF = FC, CG = \frac{1}{2}GA$. If the area of $[ABC] = 36$, find the area of the quadrilateral $DEFG$.

Problem 8.12 In the square shown in the diagram, the side length is 6, and the sum of the areas of the two shaded regions is 12. Find the area of quadrilateral $ABCD$.

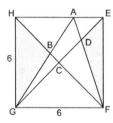

Copyright © ARETEEM INSTITUTE. All rights reserved.

8.3 Practice

Problem 8.13 A square is formed by putting 4 congruent isosceles right triangles at the corners. The shaded square is the region not covered by the triangles. Find the area of the shaded square, if

(a) As shown in the diagram below, the triangles just touch.

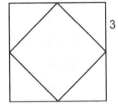

(b) As shown in the diagram below, the triangles overlap a little. (Note: In this case, 3 is *not* the side length of the isosceles triangle.)

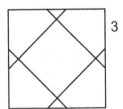

Problem 8.14 In parallelogram $ABCD$ as shown, $BC = 12$. Triangle BCE is a right triangle where $\overline{BE}$ is the hypotenuse, and $EC = 9$. If $AE = CD$, what is the area of the full figure (the pentagon $BCDFE$)?

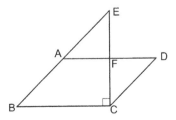

Problem 8.15 Using the above figure (not drawn to scale), if $BE = 10$, $AD = 8$ and the area of right triangle BCE is equal to the area of parallelogram $ABCD$, what is the area of trapezoid $ABCF$?

Problem 8.16 (2014 AMC 8) Six rectangles each with a common base width of 2 have lengths of $1, 4, 9, 16, 25$, and 36. What is the sum of the areas of the six rectangles?

Problem 8.17 As shown in the diagram, square $ABCD$ has side length 5. Let E and F be the midpoints of $\overline{AB}$ and $\overline{BC}$ respectively. Find the area of quadrilateral $BFGE$.

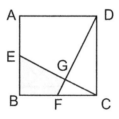

8.3 Practice

Problem 8.18 In the diagram, $\triangle ABC$ and $\triangle DEF$ are two congruent isosceles right triangles. Given that $AB = 9, EC = 3$, find the area of the shaded region.

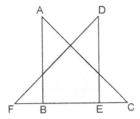

Problem 8.19 In the figure containing right triangle BCE and parallelogram $ABCD$ shown below, $EF = 3, EG = 5$ and $BG = 20$. Find the area of the shaded region.

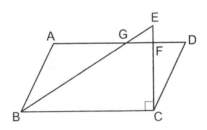

Problem 8.20 (2012 AMC 8) An equilateral triangle and a regular hexagon have equal perimeters. If the area of the triangle is 4, what is the area of the hexagon?

Problem 8.21 (2006 AMC 8) Points A, B, C and D are midpoints of the sides of the larger square. If the larger square has area 60, what is the area of the smaller square?

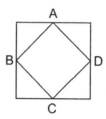

Problem 8.22 In the diagrams below, the figures are generated by taking the middle third of every side of the previous figure and forming equilateral triangles from the thirds.

(a) Let's start with the equilateral triangle! If the equilateral triangle has side length 9, what is the area of the equilateral triangle?

(b) What is the area of the second figure?

(c) What is the area of the third figure?

Copyright © ARETEEM INSTITUTE. All rights reserved.

8.3 Practice

Problem 8.23 (2003 AMC 8) In the figure, the area of square $WXYZ$ is 25 cm^2. The four smaller squares have sides 1 cm long, either parallel to or coinciding with the sides of the large square. In $\triangle ABC$, $AB = AC$, and when $\triangle ABC$ is folded over side $\overline{BC}$, point A coincides with O, the center of square $WXYZ$. What is the area of $\triangle ABC$, in square centimeters?

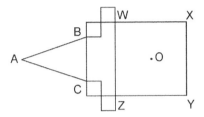

Problem 8.24 In parallelogram $ABCD$ as shown, $BC = 10$. Triangle BCE is a right triangle where $\overline{BE}$ is the hypotenuse, and $EC = 8$. Given that $[ABG] + [CDF] - [EFG] = 10$, find the length of $\overline{CF}$.

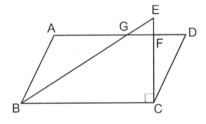

Problem 8.25 In the diagram, $\triangle ABC$ and $\triangle DEF$ are two congruent isosceles right triangles. Given that $ADFC$ is a 4×3 rectangle, find the area of the shaded region.

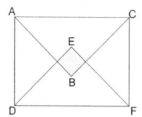

9. Circle Around

This chapter introduces the concepts and problem-solving techniques involving circles.

The concepts introduced in this chapter directly correspond to Common Core Math Standards as shown in the following table.

Grade	Standards
6th Grade	6.RP.1, 6.RP.3, 6.EE.1, 6.EE.6, 6.G.1
7th Grade	7.EE.4, 7.G.1, 7.G.4, 7.G.6
8th Grade	8.NS.2, 8.EE.2

In addition to the standards above, problems and concepts in this section will help strengthen understanding of the following domains.

Grade	Domains
6th Grade	6.RP, 6.EE, 6.G
7th Grade	7.RP, 7.EE, 7.G
8th Grade	8.NS, 8.EE, 8.G

Copyright © ARETEEM INSTITUTE. All rights reserved.

Circles

A circle is a collection of points of equal distance from a given center point. This distance is called the radius of the circle.

Chords and Diameters

- Given two points A and B on a circle, the segment $\overline{AB}$ is called a chord.
- If a chord $\overline{AB}$ contains the center of the circle, call $\overline{AB}$ a diameter.

Remark

The length of a diameter is always twice the radius of a circle. Since all diameters have the same length, this length is often refered to as the diameter of the circle.

Arcs

The portion of a circle that lies above or below a chord $\overline{AB}$ is called an arc. If the arc is more than half a circle it is called a major arc, less than half a circle is called a minor arc, and half a circle is called a semicircle. An arc from A to B will be denoted $\widehat{AB}$.

Remark

Given two points A and B on a circle, there can be some confusion about the meaning of $\overline{AB}$ as seen in the diagram below.

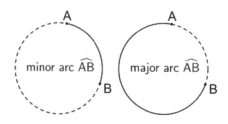

Copyright © ARETEEM INSTITUTE. All rights reserved.

From context the meaning of $\overset{\frown}{AB}$ is usually clear, but if clarification is needed, the arc can be labelled as either minor or major.

Central Angles and Sectors

Suppose we connect the ends of an arc to the center of the circle as in the diagram below.

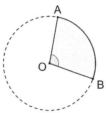

- $\angle AOB$ is called a central angle as it contains the center of the circle.
- The angular size of the arc $\overset{\frown}{AB}$ is equal to the central angle $\angle AOB$.
- The shaded region inside the arc and the two radii is called a sector.

Remark

A full circle is $360°$, a half circle (or semicircle) is $180°$, and a quarter circle is $90°$.

Example 9.1

The diagram below can be useful for reference to help understand the above terms.

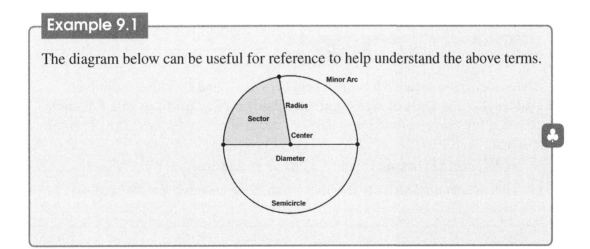

Now that we have reviewed the basic vocabulary for studying circles and other circular shapes, we can turn our attention to measurements in circles.

Copyright © ARETEEM INSTITUTE. All rights reserved.

> **Pi or π**
>
> A very useful number in geometry is the number pi or π, which is defined to be the ratio between the circumference of a circle and its diameter. It is also used to calculate the area and perimeter of circles.
>
> $$\pi = 3.14159265358979323846\cdots$$
>
> where the decimals continue forever, with no repeating patterns.
> It is common to leave answers in terms of π (for example, 5π could be a simplified answer.
> In cases where an approximation is needed, some common approximations are $\pi \approx 3.14$ or $\pi \approx \frac{22}{7}$

> **Example 9.2 Area and Perimeter of a circle**
>
> Suppose we are given a circle with radius r.
> - The area of the circle is $\pi \times r^2$.
> - The perimeter of the circle, which is also called the circumference, is $2 \times \pi \times r$.

> **Arc Lengths and Areas of Sectors**
>
> The arc length of an arc $\widehat{AB}$ is the length between A and B around the circle.
> Arc lengths and areas of sectors are best thought of as fractions of a full circle. Recall a full circle with radius r has area πr^2 and circumference $2\pi r$. Then for example
> - A semicircle has arc length $\frac{1}{2} \times 2\pi r = \pi r$ and area $\frac{1}{2} \times \pi r^2 = \frac{\pi r^2}{2}$.
> - A quarter circle has arc length $\frac{1}{4} \times 2\pi r = \frac{\pi r}{2}$ and area $\frac{1}{4} \times \pi r^2 = \frac{\pi r^2}{4}$.

Copyright © ARETEEM INSTITUTE. All rights reserved.

> **Remark**

> In general, suppose an arc $\widehat{AB}$ has angular size $\theta°$. Then the arc length of $\widehat{AB}$ is
> $$\frac{\theta°}{360°} \times 2\pi r.$$
> Similar, the sector formed by $\widehat{AB}$ has area
> $$\frac{\theta°}{360°} \times \pi r^2.$$

> **Tangent**
>
> - A line is tangent to a circle if the line intersects the circle exactly once. If this intersection point is P, we say the line is tangent to the circle at P.
> - Similarly, two circles are tangent if they intersect at exactly one point. As before, if this point is P, we will say the two circles are tangent at a point A.
> - Note: If two objects are tangent at a point P, it is often useful to think of them as "just touching" at P.

Example 9.3 Tangents are Perpendicular

Two very useful facts about tangents are summarized below.

- If a line is tangent to a circle at point P and $\overline{OP}$ is a radius of the circle, then $\overline{OP}$ is perpendicular to the line, as in the picture below:

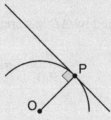

- If two circles (with centers N, O) are tangent at a point P, then the line perpendicular to $\overline{NO}$ going through P is tangent to both circles, as in the picture below:

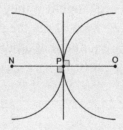

9.1 Example Questions

Example 9.4

A circular dining table has diameter 2 meters and height 1 meter. A square tablecloth is placed on the table, and the four corners of the tablecloth just touch the floor. Find the area of the tablecloth in square meters.

Solution

As shown in the diagram, the tablecloth's diagonal is 4 meters, thus the area is

$$4^2 \div 2 = 8 \text{ m}^2$$

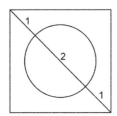

Example 9.5

The shape in the diagram consists of three semicircles with radii 5, 3, and 2, arranged as shown. Calculate the perimeter and area of the shaded region.

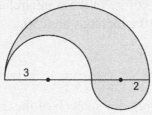

Solution

The perimeter of the figure is determined by

$$2\pi \times \frac{3}{2} + 2\pi \times \frac{2}{2} + 2\pi \times \frac{5}{2} = 10\pi,$$

and the area of the figure is determined by

$$\frac{\pi 5^2}{2} - \frac{\pi 3^2}{2} + \frac{\pi 2^2}{2} = 10\pi.$$

Example 9.6

(2000 AMC 8) Three circular arcs of radius 5 units bound the region shown.

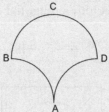

Arcs *AB* and *AD* are quarter-circles, and arc *BCD* is a semicircle. What is the area, in square units, of the region?

Solution

Note that if we draw line segments *AC* and *BD* with intersection point *O*, we can rearrange the figure so that region *AOD* fits in the top left corner of the semicircle, and region *BOC* fits in the top right corner of the semicircle. When this is done, we recreate a rectangle of dimensions 5×10 which has area 50.

Example 9.7

Find the areas of the shaded regions in each of the diagrams. Note: for the most part shapes are drawn to scale, so you may assume angles that look right are in fact 90°, etc.

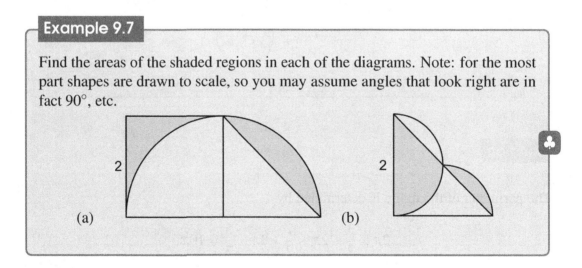

Solution

For (a), note the two shaded regions combine to form half a 2 by 2 square. Thus the area is 2.

9.1 Example Questions

For (b), note the two shaded regions combine to form a semicircle with radius 1. Thus the area is $\frac{\pi}{2}$.

Example 9.8

Find the areas of the shaded regions in each of the diagrams. Note: for the most part shapes are drawn to scale, so you may assume angles that look right are in fact 90°, etc.

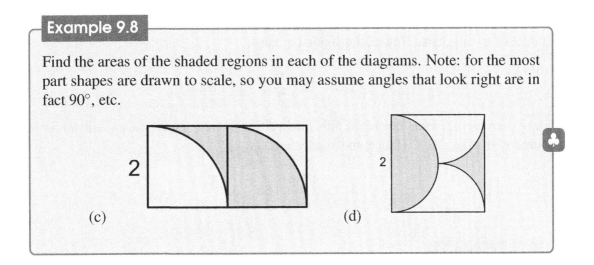

Solution

For (a), note the two shaded regions combine to form a 2 by 2 square. Thus the area is 4. For (b), note the two shaded regions combine to form a 2 by 1 rectangle. Thus the area is 2.

Example 9.9

Find the areas of the shaded regions in each of the diagrams.

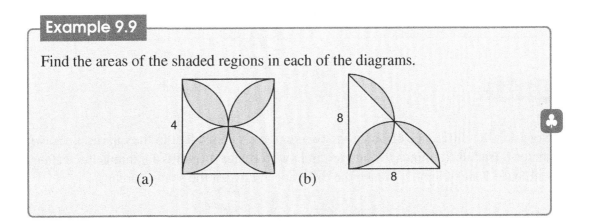

Solution

For (a), start by drawing the diagonals. Note the 8 regions can be rearranged to form 2 circles (each with radius 2) so that a square (with diagonal 4) is left unshaded in the middle of each square. Hence, the shaded area is

$$2 \times 2^2 \pi - 2 \times \frac{4^2}{2} = 8\pi - 16.$$

For (b), we use a similar idea to (a), this time getting a circle with radius 4 with a missing square with diagonal 8. Hence, the shaded area is

$$4^2 \pi - \frac{8^2}{2} = 16\pi - 32.$$

Example 9.10

In the diagram below, the two quartercircles have radii 1 and 2 respectively. Find the difference between the areas of the two shaded regions.

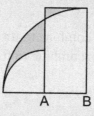

Solution

Note that the difference between the two regions is equivalent to the quartercircle with center B (radius 2), minus the quartercircle with center A (radius 1), minus the rectangle with base 1 and height 2. Hence the difference between the shaded areas is

$$\frac{2^2 \pi}{4} - \frac{1^2 \pi}{4} - 2 \times 1 = \frac{3\pi}{4} - 2.$$

Copyright © ARETEEM INSTITUTE. All rights reserved.

9.1 Example Questions

Example 9.11

(2004 AMC 8) Two 4×4 squares intersect at right angles, bisecting their intersecting sides, as shown.

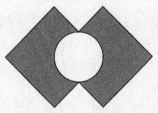

The circle's diameter is the segment between the two points of intersection. What is the area of the shaded region created by removing the circle from the squares?

Solution

If the circle was shaded in, the intersection of the two squares would be a smaller square with half the side length, 2. The area of this region would be the two larger squares minus the area of the intersection, the smaller square. This is

$$4^2 + 4^2 - 2^2 = 28.$$

Furthermore, note that the diagonal of this smaller square created by connecting the two points of intersection of the squares is the diameter of the circle. Using the side ratio property of $45 - 45 - 90$ triangles, the length of this diagonal is $2\sqrt{2}$. The radius is half the diameter, $\sqrt{2}$.

The area of the circle is

$$\pi r^2 = \pi(\sqrt{2})^2 = 2\pi.$$

Therefore, the area of the shaded region is the area of the circle minus the area of the two squares which is $28 - 2\pi$.

Example 9.12

Suppose in the diagram below the side length of the square is 4. Find the total area of the whole shape.

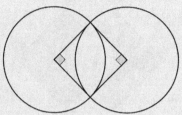

Solution

Both circles have radii 4, so the desired area is the sum of the two circles minus their overlap region. The overlap region can be calculated by adding two quartercircles and then subtracting the square. The overlap region of the two circles is

$$2 \times \frac{4^2 \times \pi}{4} - 4^2 = 8\pi - 16.$$

Hence the final answer is

$$2 \times 4^2 \times \pi - (8\pi - 16) = 24\pi + 16.$$

Example 9.13

(2008 AMC 8) Two circles that share the same center have radii 10 meters and 20 meters. An aardvark runs along the path shown, starting at A and ending at K.

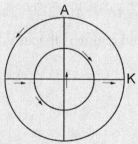

How many meters does the aardvark run?

9.1 Example Questions

Solution

We calculate the individual lengths of the circle. The first part of the path corresponds to a length of $\frac{1}{4}$ the circumference of the big circle. This is equal to

$$\frac{2\pi r}{4} = \frac{\pi r}{2} = \frac{20\pi}{2} = 10\pi.$$

The second part of the path is the difference between the big and small radii of the circles, hence has length
$$20 - 10 = 10.$$

The third part of the path corresponds to $\frac{1}{4}$ of the circumference of the small circle, which can be computed as follows:

$$\frac{\pi r}{2} = \frac{10\pi}{2} = 5\pi.$$

The fourth part is the diameter of the small circle, hence has length
$$2 \times 10 = 20.$$

The fifth part of the path has the same length as the third part, hence has length 5π. Similarly the last part of the path is the same as the second, hence has length 10. Combining all of the calculated lengths yields the final length of

$$10\pi + 10 + 5\pi + 20 + 5\pi + 10 = 20\pi + 40.$$

9.2 Quick Response Questions

Problem 9.1 What is the area of a circle with radius 5? Circumference?

Problem 9.2 In the following diagram, circle O has a radius length of 3 cm and a ring with thickness measure 1 cm is placed around circle O 1 cm apart. Is the area of circle O greater than the area of the ring?

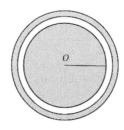

Problem 9.3 Find the circumference of a circle with area 16π.

Problem 9.4 Find the area of a circle with circumference 16π.

9.2 Quick Response Questions

Problem 9.5 Tom and Jerry each eat some pizza. Tom eats a quarter of his pizza, which has a radius of 10 inches. Jerry eats all of his pizza, which has a diameter of 10 inches.

(a) Who has eaten more crust?

(b) Who has eaten more pizza in total?

9.3 Practice

Problem 9.6 Given a semicircle, let $\overline{AB}$ be its diameter, and O be the center. Let the radius be 4. Randomly select a point C on the arc. What is the maximum possible area of $\triangle ABC$?

Problem 9.7 Find the areas of the shaded regions in each of the diagrams. Note: for the most part shapes are drawn to scale, so you may assume angles that look right are in fact 90°, etc.

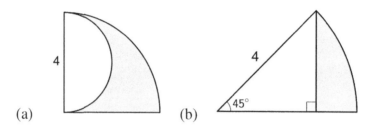

Problem 9.8 Find the areas of the shaded regions in each of the diagrams.

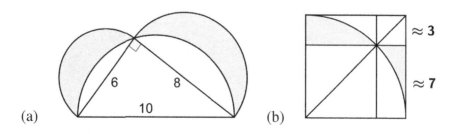

Copyright © ARETEEM INSTITUTE. All rights reserved.

Problem 9.9 Find the areas of the shaded regions in each of the diagrams.

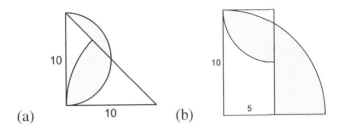

Problem 9.10 Find the areas of the shaded regions in each of the diagrams.

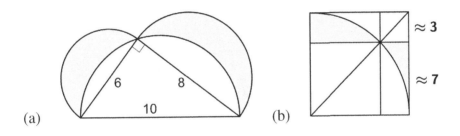

Problem 9.11 In the diagram below, the area of the shaded region is 4. What is the area of the semicircle with diameter $\overline{OA}$?

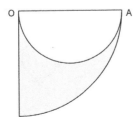

Problem 9.12 Suppose you have a triangle and a semicircle as in the diagram below. Find the difference between the area of region *A* and region *B*.

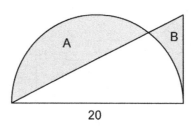

Problem 9.13 In the diagram, *ABCD* is a parallelogram, *O* is the center of the circle. Given that $[ABCD] = 8$, find the area of the shaded region $\triangle BOC$.

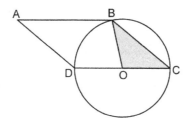

Problem 9.14 All the smaller circles in the diagram below have radii 1. Find the perimeter of the shaded region.

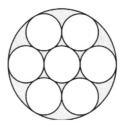

9.3 Practice

Problem 9.15 (2010 AMC 8) A decorative window is made up of a rectangle with semicircles at either end as in the diagram below.

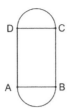

The ratio of *AD* to *AB* is 3 : 2. And *AB* is 30 inches. What is the ratio of the area of the rectangle to the combined area of the semicircles?

Problem 9.16 Find the perimeter of the shaded regions in each of the diagrams. Note: for the most part shapes are drawn to scale, so you may assume angles that look right are in fact 90°, etc.

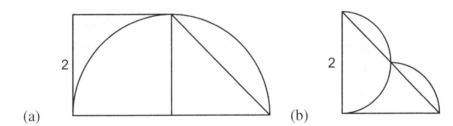

Problem 9.17 Suppose you have a circle of radius 1. A rectangle is inscribed in the circle, and a rhombus is inscribed in the rectangle, as shown in the diagram. What is the side length of the rhombus?

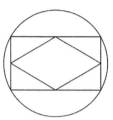

Problem 9.18 In the diagram below, the largest circle has radius 5, and the two smaller circles have radii 3 and 4 respectively. The region A is the overlapped region of the two smaller circles. Find the difference between the area of the shaded region and the area of region A.

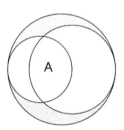

9.3 Practice

Problem 9.19 In the diagram, the area of region A equals the area of region B plus $50\pi - 100$. Find the height of the triangle.

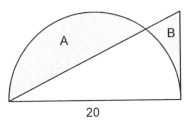

Problem 9.20 (2013 AMC 8) A 1×2 rectangle is inscribed in a semicircle with the longer side on the diameter. What is the area of the semicircle?

Problem 9.21 Congruent circles with centers A and B intersect such that AB is a radius of each circle. If $AB = 3$, what is the area of the intersecting region shared by the two circles?

Problem 9.22 Three circles having each a radius of 3 cm are tangent to each other. Consider the triangle formed by joining their three centers. What is the area of the region inside this triangle but outside the three circles.

Problem 9.23 Let *ABCD* be a square with edge length of 3 cm. We draw a circle of center *A* and radius 3 cm and a circle of center *C* and radius 3 cm. What is the ratio of the area of the circle's intersection to the area of the square?

Problem 9.24 (2010 AMC 8) The two circles pictured have the same center *C*.

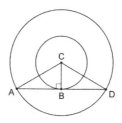

Chord $\overline{AD}$ is tangent to the inner circle at *B*, *AC* is 10, and chord $\overline{AD}$ has length 16. What is the area between the two circles?

Problem 9.25 Let *ABC* be an equilateral triangle with coordinates $A = (0,0)$ and $B = (1,0)$. Triangle *ABC* is rotated counterclockwise about point *A* so that the image of *C* lies on the *y*−axis. Find the length of the path formed by point *C* during the rotation.

10. Circle Back

We are now ready to tackle even more challenging problems involving areas and perimeters of circular shapes.

The concepts introduced in this chapter directly correspond to Common Core Math Standards as shown in the following table.

6th Grade	6.RP.1, 6.RP.3, 6.EE.1, 6.EE.6, 6.G.1
7th Grade	7.EE.4, 7.G.1, 7.G.4, 7.G.6
8th Grade	8.NS.2, 8.EE.2

In addition to the standards above, problems and concepts in this section will help strengthen understanding of the following domains.

6th Grade	6.RP, 6.EE, 6.G
7th Grade	7.RP, 7.EE, 7.G
8th Grade	8.NS, 8.EE, 8.G

Copyright © ARETEEM INSTITUTE. All rights reserved.

10.1 Example Questions

Example 10.1

In the diagram below, find the ratio between the areas of the shaded region and the big circle.

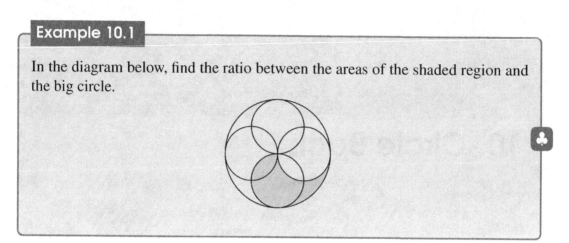

Solution

Since the sum of the areas of the smaller circles equals the area of the large circle, the overlapping regions of the smaller circle have the same area as the regions between the large circle and the small circles.

Consequently, considering the partially shaded small circle, the missing portion has the same area as the region outside this small circle.

Therefore the area of the shaded region equals the area of a small circle, which is $\frac{1}{4}$ of the large circle. So the ratio is $1 : 4$.

Example 10.2

(2012 AMC 8) A circle of radius 2 is cut into four congruent arcs. The four arcs are joined to form the star figure shown.

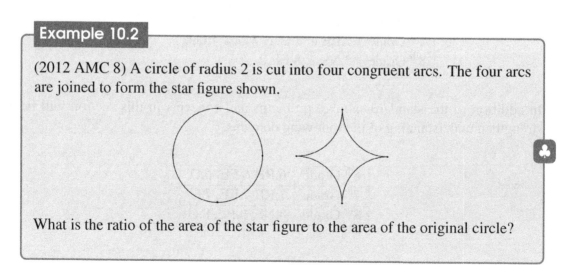

What is the ratio of the area of the star figure to the area of the original circle?

Copyright © ARETEEM INSTITUTE. All rights reserved.

10.1 Example Questions

Solution

As shown below, if we draw a square around the star figure, we note that the side length of this square is 4 since the side length is equal to the diameter of the circle.

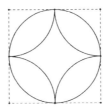

The square forms four quarter circles, each with radius 2. Therefore, the area of the star figure is
$$4^2 - 2^2\pi = 16 - 4\pi.$$
The area of the circle is 4π and the desired ratio of the two areas is
$$\frac{4-\pi}{\pi}.$$

Example 10.3

(2014 AMC 8) The circumference of the circle with center O is divided into 12 equal arcs, marked the letters A through L as seen below. What is the number of degrees in the sum of the angles x and y?

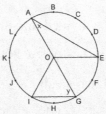

Solution

Since the points are distributed equally, the arc length between two consecutive points is
$$360 \div 12 = 30°.$$

Copyright © ARETEEM INSTITUTE. All rights reserved.

Angle *AOE* covers the arc of 120°. This implies that the angle measure of *AOE* is equal to 120°. Two of the sides of the triangle are radii of the circle. This implies that triangle *AOE* is isosceles and
$$x = \frac{180-120}{2} = 30°$$
Similarly, since angle *GOI* covers the arc of 60°, and triangle *GOI* is isosceles, we have
$$y = \frac{180-60}{2} = 60°.$$
Therefore, the sum of angle measure x and y is 90°.

Example 10.4

A dog is tied with a rope to the top corner of a building whose base is an equilateral triangle with side length 3 m, as shown in the diagram. The length of the rope is 4 m. Find the total area of the region (in m²) that the dog can reach.

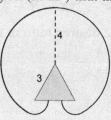

Solution

The region the dog can reach can be divided into one big
$$360° - 60° = 300°$$
sector of a circle with radius 4 and two small
$$180° - 60° = 120°$$
sectors of radius $4 - 3 = 1$. As the big and small sectors are respectively
$$\frac{300}{360} = \frac{5}{6}, \frac{120}{360} = \frac{1}{3}$$
of a circle, the area is
$$\frac{5}{6} \times 4^2 \pi + \frac{2}{3} \times 1^2 \pi = \frac{40}{3}\pi + \frac{2}{3}\pi = 14\pi.$$

10.1 Example Questions

Example 10.5

As shown in the diagram, in rectangle $ABCO$, given $[ABD] - [BCD] = 10$. Find the area of the shaded region.

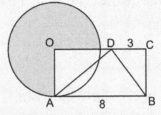

Solution

Note that $BC = AO$ and AO is the radius of the circle. Therefore, triangle ABD has height BC and base length 8. Similarly, triangle BCD also has height BC and base length 3. Therefore,

$$10 = [ABD] - [BCD] = \frac{1}{2} \times 8 \times BC - \frac{1}{2} \times 3 \times BC = \frac{5BC}{2}.$$

Solving the above equation yields $BC = 4$, which is the radius of the circle. The shaded region has area

$$\frac{3}{4} \times 4^2 \pi = 12\pi.$$

Example 10.6

(2014 AMC 8) Rectangle $ABCD$ has sides $CD = 3$ and $DA = 5$. A circle of radius 1 is centered at A, a circle of radius 2 is centered at B, and a circle of radius 3 is centered at C.

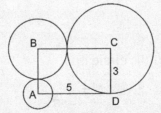

To the nearest 0.5 square units, what is the area of the region inside the rectangle but outside all three circles?

Solution

Note that the area in the rectangle but outside the circles is the area of the rectangle minus the area of all three of the quarter circles in the rectangle. It is plain to see that the area of the rectangle is
$$3 \times 5 = 15.$$
The area of all 3 quarter circles is
$$\frac{\pi}{4} + \frac{\pi(2)^2}{4} + \frac{\pi(3)^2}{4} = \frac{14\pi}{4} = \frac{7\pi}{2}.$$
Therefore, the area in the rectangle but outside the circles is $15 - \frac{7\pi}{2}$. Using the approximation $\pi \approx \frac{22}{7}$, we get
$$15 - \frac{7}{2} \times \frac{22}{7} = 15 - 11 = 4.$$

Example 10.7

In the two diagrams given below, the lengths of segments $\overline{AB}$ are both equal to 2. Further suppose the inside circle in the second diagram has radius 1. Which shaded region has a larger area?

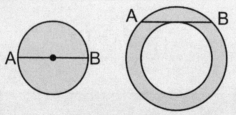

Solution

In the first diagram, the area of the circle is
$$\frac{2^2 \pi}{4} = \pi.$$

In the second diagram shown below, let O be the center of the circles, and C be the midpoint of $\overline{AB}$.

10.1 Example Questions 213

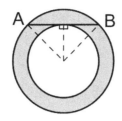

Then $\overline{OC} \perp \overline{AB}$, so
$$OA^2 = 1^2 + 1^2 = 2.$$

Hence, the area of the shaded region is
$$2\pi - \pi = \pi.$$

Example 10.8

Three circles all have radii 2, and each passes through the centers of the other two, as shown in the diagram. Find the area of the shaded region.

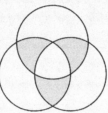

Solution

Create four equilateral triangles as shown below.

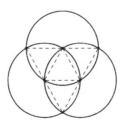

If we regroup the shaded area, we can create three circular sectors formed by a 60° angle. Note that,
$$\frac{60}{360} = \frac{1}{6},$$
implies that the circular sectors each represent $\frac{1}{6}$ of the circle with radius 2. Therefore, combining the three parts yields a semicircle. The area of the shaded region is
$$\frac{2^2 \pi}{2} = 2\pi.$$

Example 10.9

(2014 AMC 8) A straight one-mile stretch of highway, 40 feet wide, is closed. Robert rides his bike on a path composed of semicircles as shown. If he rides at 5 miles per hour, how many hours will it take to cover the one-mile stretch? Note: 1 mile = 5280 feet

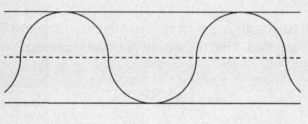

Solution

Since the entire road is 40 feet wide, each lane is 20 feet wide which indicates that Robert is riding paths in semicircles of radius 20 feet. Given that the road is 1 mile long, Robert completes
$$\frac{5280}{40} = 132$$
semicircles total in his path. The distance covered by Robert is computed by determining the length of each semicircle with radius 20. This length is
$$\frac{1}{2}\pi(2 \times 20) = 20\pi$$
and therefore, the length that Robert covers in a one-mile increment on the road is
$$132 \times 20\pi = 2640\pi$$
feet or $\frac{\pi}{2}$ miles. At 5 miles per hour, it takes him $\frac{\pi}{10}$ hours to complete the path.

Copyright © ARETEEM INSTITUTE. All rights reserved.

10.1 Example Questions

Example 10.10

In the diagram, the area of the shaded region is 25. Find the area of the annular region (the region between the two circles).

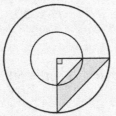

Solution

Let R be the bigger radius and r be the smaller radius. Since the radii r and R form two $45-45-90$ triangles, we can determine that the short base of the trapezoid has length

$$\sqrt{r^2 + r^2} = r\sqrt{2}$$

and the long base of the trapezoid has length

$$\sqrt{R^2 + R^2} = R\sqrt{2}.$$

The height of the trapezoid can further be indicated by creating an additional right triangle shown below:

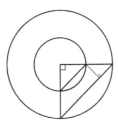

Note that the difference of the radii is equal to the hypotenuse of our newly formed $45-45-90$ triangle. Therefore, the height of the trapezoid is

$$(R-r) \div \sqrt{2} = \frac{R-r}{\sqrt{2}}$$

Then
$$\frac{1}{2} \times \frac{R-r}{\sqrt{2}} \times (r\sqrt{2}+R\sqrt{2}) = \frac{1}{2}(R^2-r^2) = 25.$$

Therefore, we have
$$R^2 - r^2 = 50,$$
and the annular region's area is
$$(R^2-r^2)\pi = 50\pi.$$

10.2 Quick Response Questions

Problem 10.1 Points A and B lie on a circle centered at point O with radius length 9 in. Suppose that the measure of $\angle AOB$ is $120°$. What is the length of arc AB?

Problem 10.2 Points A and B lie on a circle centered at point O with radius length 10 in. Suppose that the area measure of the sector formed by $\angle AOB$ is 20π in^2. What is $\angle AOB$?

Problem 10.3 Find the areas of the shaded regions in each of the diagrams below. Note: for the most part shapes are drawn to scale, so you may assume angles that look right are in fact $90°$, etc.

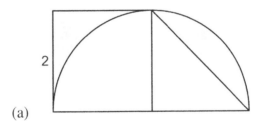

(a)

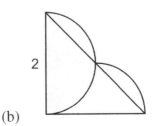

(b)

Problem 10.4 A circular dining table has diameter 2 meters and height 1 meter. A square tablecloth is placed on the table, and the four corners of the tablecloth just touch the floor. Find the area of the tablecloth in square meters.

Problem 10.5 The shape in the diagram consists of three semicircles with radii 5, 3, and 2, arranged as shown. Calculate the perimeter and area of the shaded region.

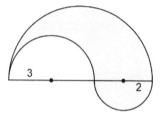

10.3 Practice

Problem 10.6 In the diagram below, the 4 smaller circles are all congruent and pass through a common point, and are all internally tangent to the bigger circle. Find the ratio between the perimeters of the shaded region and the bigger circle.

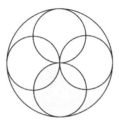

Problem 10.7 A sheep is tied with a rope to the upper-left corner of the square barn on the grass field. The length and width of the barn are 10, as shown in the diagram, and the length of the rope is 20. Find the area of the region that the sheep can reach.

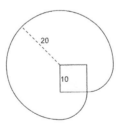

Problem 10.8 In the diagram, the radius of the circle is 4. If we are also given that the area of the shaded region is 14π, find the area of the triangle.

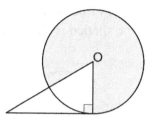

Note: This diagram does not reflect the actual shape: the angle of the unshaded sector is *not* $60°$.

Problem 10.9 Six cylindrical pencils are tied together with a rubber band. A cross section of it is shown in the diagram. The radius of the pencils is 1. Find the current length of the rubber band.

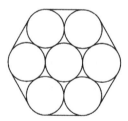

10.3 Practice

Problem 10.10 Equilateral triangle *ABC* pictured below has side length 1. It is placed on a straight line at position I as shown. Roll the triangle along the line about vertex *C* to reach position II, and roll again to reach position III. Find the total length of the path that vertex *A* traveled.

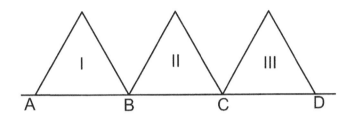

Problem 10.11 (2010 AMC 8) Semicircles *POQ* and *ROS* pass through the center *O*.

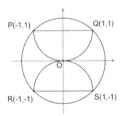

What is the ratio of the combined areas of the two semicircles to the area of circle *O*?

Problem 10.12 In the diagram below, the area of the shaded region (area between the two squares) is 200. What is the area of the annular region (between the two circles)?

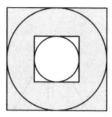

Problem 10.13 Given two circles, whose perimeters have ratio $3 : 2$. Also given that the difference between their areas is 10 cm^2. What is the sum of their areas (in cm^2)?

Problem 10.14 Rectangle I has length 15 and width 8, and is placed on a straight line as shown in the diagram. Roll the rectangle about vertex B by $90°$ to reach position II. Then roll about vertex C, and so on, until the original vertex A reaches point E. Calculate the total length of the path that vertex A traveled.

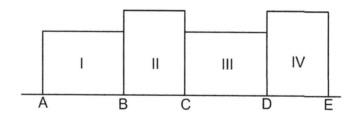

Copyright © ARETEEM INSTITUTE. All rights reserved.

10.3 Practice

Problem 10.15 (2008 AMC 8) Margie's winning art design is shown.

The smallest circle has radius 2 inches, with each successive circle's radius increasing by 2 inches. Approximately what percent of the design is black?

Problem 10.16 Find the perimeter of the shaded regions in each of the diagrams. Note: for the most part shapes are drawn to scale, so you may assume angles that look right are in fact $90°$, etc.

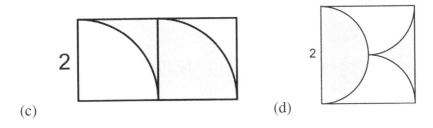

(c)   (d)

Problem 10.17 Find the perimeter of the shaded regions in each of the diagrams.

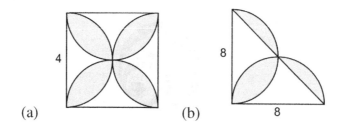

(a) (b)

Problem 10.18 (2011 AMC 8) A circle with radius 1 is inscribed in a square and circumscribed about another square as shown.

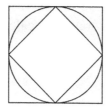

What is the ratio of the circle's shaded area to the area between the two squares?

Problem 10.19 Congruent circles centered at O and P are externally tangent, and a line through P is tangent to O at A. If $AP = 15$, what is the length of the radius of circle P?

10.3 Practice

Problem 10.20 Let A be a circle with radius 1. Let B be the inscribed regular hexagon inside circle A. Let C be the inscribed circle inside hexagon B. What is the ratio of the area of circle C to the area of circle A?

Problem 10.21 Three congruent circles with centers A, B, and C intersect such that AB, AC, and BC are radii of two circles. If $AB = 3$, what is the area of the intersecting region shared by the three circles?

Problem 10.22 Rectangle I has length 4 and width 3, and is placed on a straight line as shown in the diagram. Roll the rectangle about vertex B by $90°$ to reach position II. Then roll about vertex C, and so on, until the original vertex A reaches point E. Calculate the total length of the path that vertex A traveled.

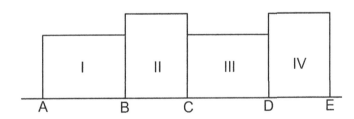

Problem 10.23 Using the above problem, find the area of the region formed by the path of vertex A.

Problem 10.24 Given a semicircle with diameter $\overline{AB}$, where $AB = 3$. Rotate this semicircle about point A by $60°$ counterclockwise, so that the point B reaches B', as shown in the diagram. Find the area of the shaded region.

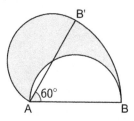

Problem 10.25 Using the same figure above, find the perimeter of the shaded region.

11. A Whole New Dimension

The world around us exists in three dimensions. There are many ways to think about three dimensions. Some common ones include:

- length, width, height
- up/down, left/right, forward/backward
- x-axis, y-axis, z-axis

This chapter focus on geometric objects in these three dimensions.

The concepts introduced in this chapter directly correspond to Common Core Math Standards as shown in the following table.

6$^\text{th}$ Grade	6.RP.1, 6.RP.3, 6.EE.1, 6.EE.6, 6.G.2, 6.G.4
7$^\text{th}$ Grade	7.EE.4, 7.G.1, 7.G.3, 7.G.6
8$^\text{th}$ Grade	8.EE.3, 8.G.1, 8.G.9

In addition to the standards above, problems and concepts in this section will help strengthen understanding of the following domains.

Copyright © ARETEEM INSTITUTE. All rights reserved.

6th Grade	6.RP, 6.EE, 6.G
7th Grade	7.RP, 7.EE, 7.G
8th Grade	8.NS, 8.EE, 8.G

Objects in Three Dimensions

- Points are zero-dimensional objects.
- Lines, line segments, and rays are all one-dimensional objects.
- If we extend a piece of paper forever, we get a two-dimensional object called a plane. All the shapes we have discussed throughout this book can be thought of as being parts of a plane.
- A three-dimensional object is called a solid. Objects such as cubes, boxes, balls, etc. are all solids.

Example 11.1 Parallel Planes

Two different planes may or may not intersect. If they intersect, they do so in a line. If they do not intersect we call the planes parallel.

Remark

Notice the similarity to lines in two dimensions.

Example 11.2 Lines in Three Dimensions

In three dimensions, two different lines either:
1. intersect at a point.
2. point in the same direction and never intersect. In this case we say the lines are parallel.
3. are not parallel, but also never intersect. In this case we say the lines are skew. For example, start with two lines that intersect in a plane and then "lift one line up" so they no longer intersect.

Copyright © ARETEEM INSTITUTE. All rights reserved.

Example 11.3 Lines and Angles

If two lines intersect in three dimensions, then there is a unique plane containing these two lines. Therefore, we can measure the angles between these lines as we have in the previous chapters.

Volume

The amount of space a solid takes up is called its volume.

Surface Area

The area on the outside of a solid is called its surface area.

Vertices, Edges, Faces

Consider the cube below:

- The 8 corner points of the cube are examples of vertices.
- The 12 line segments of the cube are examples of edges.
- The 6 squares on the surface of the cube are examples of faces.

Unit Cube

A cube with dimensions one unit by one unit by one unit has area one cubic unit.

Example 11.4 Cube

Suppose a cube has side length s.
- The cube has 8 vertices, 12 edges, and 6 faces. The faces are all squares.
- The cube has volume s^3.
- The cube has surface area $6s^2$.

Example 11.5 Rectangular Prism

A rectangular prism or box is a solid as in the diagram below.

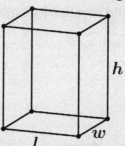

Suppose the box above has length l, width w, and height h.
- The box has 8 vertices, 12 edges, and 6 faces. The faces are all rectangles.
- The box has volume $l \times w \times h$.
- The box has surface area $2(l \times w + l \times h + w \times h)$.

Example 11.6 Sphere

A perfectly round solid is called a sphere or a ball. Similar to a circle, the distance from the outside to the center is called the radius, as in the diagram below.

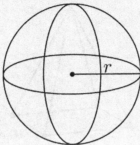

Suppose the sphere above has radius r.
- The sphere has volume $\frac{4}{3}\pi r^3$.
- The sphere has surface area $4\pi r^2$.

Example 11.7 Cylinder

A solid shaped like a can is called a cylinder. A cylinder has a top and bottom called a base that is a circle, as in the diagram below.

Suppose the cylinder above has radius r and height h.
- The cylinder has volume $\pi r^2 \times h$.
- The cylinder has surface area $2\pi r \times h + 2\pi r^2$.

Example 11.8 Rectangular Pyramid

Consider the solid shown below, called a rectangular pyramid, with a rectangular base connected to a single point, called the apex, above the center of the base.

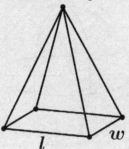

Suppose the pyramid above has a rectangular base with length l and width w. Further suppose that the height of the pyramid is h.

- The pyramid has 5 vertices, 8 edges, and 5 faces. The bottom face is a rectangle, while the other four faces are triangles.
- The pyramid has volume $\frac{1}{3} l \times w \times h$.
- The pyramid's surface area can be calculated by finding the area of each face separately and adding them together.

Example 11.9 Cone

Consider the solid shown below, called a cone, with a circular base connected to a single point, called the apex, above the center of the base.

Suppose the pyramid above has a circular base with radius r. Further suppose that the height of the pyramid is h.

- The cone has volume $\frac{1}{3}\pi r^2 \times h$.
- The cone has surface area $\pi r^2 + \pi r \times \sqrt{r^2 + h^2}$.

Remark

Note that a rectangular pyramid fits inside a rectangular prism, while a cone fits inside a cylinder, as shown in the diagram below.

 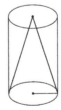

In fact, the pyramid or cone has $\frac{1}{3}$ of the volume of the full shape (either the prism or the cylinder). This can be very helpful when remembering formulas.

11.1 Example Questions

Example 11.10

Suppose you have a box (rectangular prism) that is 2 feet long, 1 foot wide, and has a height of 6 inches. Suppose you want to double the volume of the box by changing one of the dimensions of the box. What are the possible new surface areas (measured in square feet)?

Solution

We can calculate that the original box has volume

$$2 \times 1 \times \frac{1}{2} = 2 \text{ft}^3.$$

Since we are interested in determining the dimensions of the box after its volume was doubled after changing the length of one side, we observe that the possible dimensions of the new figure are $4 \times 1 \times .5$ or $2 \times 2 \times .5$ or $2 \times 1 \times 1$ if we are restricted to only changing one of the sides of the prism.

Therefore, the surface areas can be computed as follows:

$$2(4 \times 1 + 4 \times .5 + 1 \times .5) = 13$$
$$2(2 \times 2 + 2 \times .5 + 2 \times .5) = 12$$
$$2(2 \times 1 + 2 \times 1 + 1 \times 1) = 10$$

These are the three possible surface areas of the rectangular prism after its area is doubled.

Example 11.11

Suppose we have a ball with radius 6. Suppose you cut the ball in half. Find the volume and surface area of the half-ball.

Copyright © ARETEEM INSTITUTE. All rights reserved.

11.1 Example Questions

Solution

First note that the full ball has volume

$$\frac{4}{3}\pi \times 6^3 = 228\pi$$

and surface area

$$4\pi \times 6^2 = 144\pi.$$

The half-ball has half the volume of the full ball, so the volume is

$$228\pi \div 2 = 114\pi.$$

For the surface area, we have half of the original surface area

$$(144\pi \div 2 = 72\pi)$$

but we *also* have an extra circle of radius 6, with area

$$\pi 6^2 = 36\pi.$$

Hence the total surface area of the half-ball is

$$72\pi + 36\pi = 108\pi.$$

Example 11.12

Suppose you have a cube with side length 1. Label the vertices of the square $ABCD - EFGH$ where $ABCD$ forms the "bottom" square and $EFGH$ forms the "upper" square as in the diagram below.

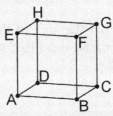

Find the distance from A to B, from A to F, and from A to G.

Solution

It is clear the distance from A to B is 1. Note that △ABF is a right triangle, so the distance from A to F is

$$\sqrt{1^2 + 1^2} = \sqrt{2}$$

using the Pythagorean theorem. We also have that △AFG is a right triangle, so using the Pythagorean theorem again we have the distance from A to G is

$$\sqrt{(\sqrt{2})^2 + 1^2)} = \sqrt{3}.$$

Note: Unsimplifying our answer, the distance from A to G is

$$\sqrt{1^2 + 1^2 + 1^2} = \sqrt{3}.$$

Example 11.13

Suppose you want to build a can of soda to hold 12π cubic inches of soda. If you want the diameter of the can to be 8 inches, how tall should you make the can? If instead you want the height of the can to be 8 inches, how wide should you make the can?

Solution

If the diameter is 8 inches, the radius is 4 inches. Therefore the base has area

$$4^2\pi = 16\pi$$

square inches. Thus, the height must be 2 inches.

If the height is 8 inches, we see that the area of the base must be 4π square inches. Therefore, the radius of the circular base must be 2 inches, so the diameter is 4 inches.

Example 11.14

Suppose you have a square pyramid with base of side length 4 and height 5. What is the volume and surface area of the pyramid?

Copyright © ARETEEM INSTITUTE. All rights reserved.

11.1 Example Questions

Solution

Using the formula for the volume, we see the pyramid has volume
$$\frac{1}{3} \times 4^2 \times 5 = \frac{80}{3}.$$
For the surface area, first note the square base has area
$$4^2 = 16.$$
The four triangular faces of the pyramid are congruent, each with a base of 4, but we need to calculate the height of these triangles. (Note this height is different than the height of the pyramid, and is referred to as the "slant height" of the pyramid.) Consider the diagram below with the slant height L.

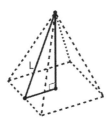

Note that the midpoint of a side, the center of the square, and the apex form a right triangle whose hypotenuse is the slant height. The other sides of this triangle are 2 (half the side length of the square) and 5 (height of the pyramid), so
$$L^2 = 2^2 + 5^2$$
so $L = \sqrt{29}$. Hence each triangle has base 4 and height $\sqrt{29}$ so each triangle has area $2\sqrt{29}$, for a total surface area of $4 \times 2\sqrt{29}$ for the triangular faces. Hence the total surface area is $16 + 8\sqrt{29}$ adding the square base.

Example 11.15

Suppose you have an ice cream cone with radius 2 inches and height 4 inches. The cone starts full of ice cream (but there is not ice cream outside the cone). After you've eaten some ice cream and some of the cone you are left with a cone with a radius and a height of 2 inches. What fraction of the ice cream have you eaten?

Solution

First note that using the volume formula for the cone, there is

$$\frac{\pi}{3} \times 2^2 \times 4 = \frac{16\pi}{3}$$

cubic inches of ice cream before any is eaten. After you've eaten some ice cream, consider the following side view (where ABC is the original cone, and ADE is the "half cone").

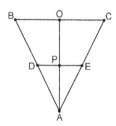

Note that $\triangle AOC \sim \triangle APE$ as they are both right triangles that share angle $\angle OAC$. Since $OA = 4, PA = 2$, we have that the ratio of sides is 2, so

$$PE = \frac{1}{2}OC = 1.$$

Hence the new cone has radius 1, and thus volume

$$\frac{\pi}{3} \times 1^2 \times 2 = \frac{2\pi}{3}$$

cubic inches of ice cream. Hence you are left with 1/8th of the ice cream you started with, so you have eaten 7/8th of the ice cream.

Example 11.16

(ZIML 2016) Billy has four cubes with side lengths 1, 2, 3, and 4 inches. He likes to form a tower stacking the cubes from largest on the bottom to smallest on the top. If the cubes are stacked in this way, what is the volume of the smallest box that can store the cubes stacked in a tower? Give your answer in inches^3.

Copyright © ARETEEM INSTITUTE. All rights reserved.

11.1 Example Questions

Solution

Since the bottom of the tower is the largest cube with side length 4, the base of the box must be 4 inches by 4 inches.

To store all the cubes vertically, the height of the box must be

$$4+3+2+1 = 10$$

inches tall. Therefore the box is 4 inches by 4 inches by 10 inches so has volume

$$4 \times 4 \times 10 = 160$$

cubic inches.

Example 11.17

In the figure below, the middle $1 \times 1 \times 1$ cube is removed from the $3 \times 3 \times 3$ cube. If we require the path to be strictly contained in the figure, find the length of the shortest path from point A to point B.

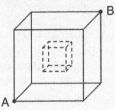

Solution

The shortest path can be determined as follows: let point C be the midpoint of one of the edges of the middle cube.

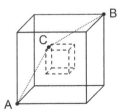

The length of AC is equivalent to determining the length of the longest diagonal of a $1 \times \frac{3}{2} \times 2$ rectangular prism and the length of CB is equivalent to determining the length of the longest diagonal of a $2 \times \frac{3}{2} \times 1$ rectangular prism. Therefore,

$$AC = \sqrt{1^2 + (\frac{3}{2})^2 + 2^2} = \frac{\sqrt{29}}{2}$$

$$CB = \sqrt{2^2 + (\frac{3}{2})^2 + 1^2} = \frac{\sqrt{29}}{2}$$

The shortest length of a path from point A to B is $\sqrt{29}$.

Example 11.18

Given a unit sphere (a unit sphere has radius 1), slice the top portion of the sphere so that we create a circle of diameter 1. Let $\overline{AB}$ be the diameter of this circle. What is the length of the shortest path from A to B along the surface of the sphere?

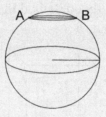

Solution

Given that the sphere has radius 1, if we define the center of the sphere to be point O, triangle AOB is an equilateral triangle with side length 1.

Therefore, the shortest path from A to B along the surface of the sphere is the $60°$ arc of a circle with radius 1.

The answer is $\frac{60}{360} \times 2\pi \times 1 = \frac{\pi}{3}$.

11.1 Example Questions

Example 11.19

An obtuse triangle with dimensions 9, 10, and 17 is rotated about the smallest side so that it creates a three-dimensional solid shown below. Determine the surface area of the solid.

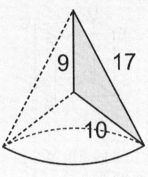

Solution

Note that in the figure above, there are two cones sharing the same circular base.

Let r be the radius of the cones and let h be the height of the smaller cone. Therefore, h and r satisfies $h^2 + r^2 = 10^2$ and $(h+9)^2 + r^2 = 17^2$.

Subtracting the first equation from the second equation,

$$(h+9)^2 + r^2 - h^2 - r^2 = 17^2 - 10^2.$$

Simplifying,
$$18h + 81 = 189,$$

therefore $h = 6$ and then $r = 8$. Thus the radius of both cones is 8, the inner cone has height 6, and the outer cone has height 15.

Therefore, the surface area of the new figure is

$$\pi \times (8 \times 17) + \pi \times (8 \times 10) = 216\pi.$$

11.2 Quick Response Questions

Problem 11.1 Recall that in a cube such as the one in the diagram below,

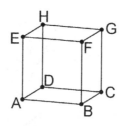

there are a total of 12 edges. Consider $\overline{AB}$. Which of the other edges intersect $\overline{AB}$? Which are parallel to $\overline{AB}$? Which are skew to $\overline{AB}$?

Problem 11.2 Suppose you have a box (rectangular prism) that is 2 feet long, 1 foot wide, and has a height of 6 inches. (1 foot equals 12 inches)

(a) How much space is inside the box? That is, what is the volume of the box?

(b) What is the surface area of the box?

Problem 11.3 Suppose we have a sphere with radius 6.

(a) Find the volume of the sphere.

11.2 Quick Response Questions

(b) Find the surface area of the sphere.

Problem 11.4 Find the volume and surface area of a ball with radius 3.

Problem 11.5 Find the volume and surface area of a square pyramid with base length 6 and height 4.

11.3 Practice

Problem 11.6 Compare the volume and surface area of a cube with side length 2 to a rectangular prism with dimensions 1, 2, 4.

Problem 11.7 Compare the volume and surface area of a cube with side length 10 to a rectangular prism with dimensions 10, 15, 6.

Problem 11.8 Find the distance from one corner to the opposite corner (for example the lower front left vertex to the upper back right vertex) of a rectangular prism with dimensions 3, 4, 12. That is, find the length of the line connecting the two corners.

Problem 11.9 Extending the previous problem, find the distance from one corner to the opposite corner of a $l \times w \times h$ rectangular prism. Note this problem can be thought of as an extension of the Pythagorean theorem.

Problem 11.10 Suppose a ball has volume $\frac{32\pi}{3}$. Find the surface area of the ball.

Copyright © ARETEEM INSTITUTE. All rights reserved.

11.3 Practice

Problem 11.11 Compare the volume of a ball with radius 4 to the combined volumes of two balls of radius 3.

Problem 11.12 Compare the surface area of a ball with radius 4 to the combined surface areas of two balls of radius 3.

Problem 11.13 How many total pairs of parallel edges does a cube contain?

Problem 11.14 Consider a $8\frac{1}{2} \times 11$ sheet of paper. One can observe that there are two ways of curling the sheet of paper so that it resembles a cylinder. Find the volumes of all possible cylinders formed by a sheet of $8\frac{1}{2} \times 11$ paper.

Problem 11.15 It is well known that about 70% of the Earth's surface is covered in water. If a sphere of radius 2 cm is painted in blue for water and green for land to create a miniature globe, find the area of the surface that is painted blue.

Problem 11.16 In the $2 \times 2 \times 2$ cubic figure below, if the path is required to be along the surface of the cube, what is the length of the shortest path from point A to B?

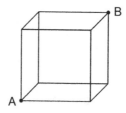

Problem 11.17 (2014 AMC 8) A cube with 3-inch edges is to be constructed from 27 smaller cubes with 1-inch edges. Twenty-one of the cubes are colored red and 6 are colored white. If the 3-inch cube is constructed to have the smallest possible white surface area showing, what fraction of the surface area is white?

Problem 11.18 (2013 AMC 8) Isabella uses one-foot cubical blocks to build a rectangular fort that is 12 feet long, 10 feet wide, and 5 feet high. The floor and the four walls are all one foot thick.

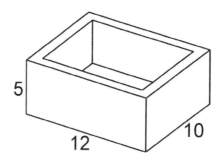

How many blocks does the fort contain?

11.3 Practice

Problem 11.19 Suppose a rectangular block of wood with dimensions 5 cm ×6 cm ×5 cm costs $50. In dollars, what is the fair price for a rectangular block of the same type of wood with dimensions 15 cm by 30 cm by 40 cm if the price is determined solely by volume?

Problem 11.20 An obtuse triangle with dimensions 9, 10, and 17 is rotated about the smallest side so that it creates a three-dimensional solid shown below. Determine the volume of the solid.

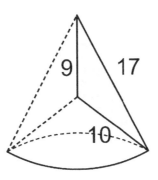

Problem 11.21 The square pyramid J_1 is a regular square pyramid composed of four equilateral trianglular faces and a square base. Given that the surface area of J_1 is $1+\sqrt{3}$, find the volume of J_1.

Problem 11.22 A regular square pyramid is placed in a cube so that the base of the pyramid and the base of the cube coincide. The vertex of the pyramid lies on the face of the cube opposite to the base. Suppose that the side length of the cube is 2 in. What is the positive difference of the surface area of the cube and the surface area of the pyramid?

Problem 11.23 A cube is increased to form a new cube so that the surface area of the new cube is 4 times that of the original cube. By what factor is the volume of the cube increased?

Problem 11.24 The mineral pyrite is commonly known as "fool's gold" due to its deceptive appearance!

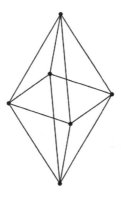

In your quest, you have uncovered a gem-shaped pyrite mineral containing a square-base located at the middle of the gem as shown above. As you fiddle with your newly uncovered gem, you note that the gem is measured to be 8 cm tall and 6 cm wide. If you plan on painting the entire gem gold, how much of the surface will be painted by gold paint?

Copyright © ARETEEM INSTITUTE. All rights reserved.

Problem 11.25 A box shaped treasure chest has outer dimensions 15 cm by 15 cm by 10 cm and the chest is 2 cm thick on all sides. Suppose you have seven dragon balls, each with diameter 5 cm, and you are interested in putting all of the dragon balls in this chest to keep them away from your enemies. Is the volume of the chest greater than the volume of the 7 dragon balls combined?

Answer Key to Problems

The answers to the quick response and practice problems in all the chapters are provided here. Only short answers are included. Proof problems are omitted below. Full solutions for all problems can be found in the solutions manual, *Geometry Problem Solving for Middle School Solutions Manual*.

Chapter 1. Counting Through Patterns Answer Key

Problem 1.1 7 points; 3 lines

Problem 1.2 A, B, F; F; D or G

Problem 1.3 m

Problem 1.4 $\overleftrightarrow{AB}, \overleftrightarrow{AF}, \overleftrightarrow{BF}$

Problem 1.5 a) 5; b) 13

Copyright © ARETEEM INSTITUTE. All rights reserved.

Problem 1.6 54°

Problem 1.7 8

Problem 1.8 5

Problem 1.9 10

Problem 1.10 54°

Problem 1.11 a) 44, b) 63

Problem 1.12 30

Problem 1.13 14

Problem 1.14 3 : 5

Problem 1.15 7

Problem 1.16 35

Problem 1.17 22

Problem 1.18 5

Problem 1.19 325 cm^2

Problem 1.20 20 cm

Problem 1.21 $\dfrac{4}{9}$

Problem 1.22 60 cm

Copyright © ARETEEM INSTITUTE. All rights reserved.

Answer Key to Problems

Problem 1.23 20

Problem 1.24 40 of each

Problem 1.25 32

Chapter 2. Practice with Measurements

Problem 2.1 44

Problem 2.2 14

Problem 2.3 30

Problem 2.4 2

Problem 2.5 24

Problem 2.6 10 in and 15 in

Problem 2.7 143 cm^2

Problem 2.8 26 cm

Problem 2.9 16

Problem 2.10 $\frac{5}{18}$

Problem 2.11 24 cm^2

Problem 2.12 49

Problem 2.13 24 in^2

Problem 2.14 20 in and 15 in

Problem 2.15 99%

Problem 2.16 44 cm

Problem 2.17 20

Problem 2.18 20%

Problem 2.19 6 cm^2

Problem 2.20 32

Problem 2.21 88

Problem 2.22 6 and 2

Problem 2.23 3600

Problem 2.24 4

Problem 2.25 7 : 32

Chapter 3. A Dance with Angles

Problem 3.1   (a) right: 90°
  (b) acute: 45°
  (c) straight: 180°
  (d) reflex: 270°

Answer Key to Problems

(e) obtuse: 150°

(f) reflex: 210°

Problem 3.2 Corresponding angles: $\angle 1 = \angle 5$; $\angle 2 = \angle 6$; $\angle 3 = \angle 7$; $\angle 4 = \angle 8$ Vertical angles: $\angle 1 = \angle 3$; $\angle 2 = \angle 4$; $\angle 5 = \angle 7$; $\angle 6 = \angle 8$

Problem 3.3   (a) $\angle 3 = \angle 5$; $\angle 4 = \angle 6$

(b) $\angle 1 = \angle 7$; $\angle 2 = \angle 8$

Problem 3.4   (a) $\angle 4 + \angle 5 = 180°$; $\angle 3 + \angle 6 = 180°$

(b) $\angle 1 + \angle 8 = 180°$; $\angle 2 + \angle 7 = 180°$

Problem 3.5 $\angle 1 = \angle 3 = \angle 5 = \angle 7 = 42°$ and $\angle 2 = \angle 4 = \angle 6 = \angle 8 = 138°$

Problem 3.6 105°

Problem 3.7 60°.

Problem 3.8 45°.

Problem 3.9 105°

Problem 3.10 115°.

Problem 3.11 80°

Problem 3.12 20°

Problem 3.13 144°

Problem 3.14 72°

Problem 3.15 55°

Copyright © ARETEEM INSTITUTE. All rights reserved.

Problem 3.16 50°

Problem 3.17 60°

Problem 3.18 60°

Problem 3.19 70°

Problem 3.20 140°

Problem 3.21 $a = 58°, b = 26°$

Problem 3.22 $c = 102°, d = 78°$

Problem 3.23 $e = 122°, f = 58°$

Problem 3.24 87°

Problem 3.25 $\dfrac{23}{7}$

Chapter 4. Magic in Triangles

Problem 4.1 (a) Answers may vary. Draw your own! Here is one example:

(b) Answers may vary. Draw your own! Here is one example:

Answer Key to Problems

(c) Answers may vary. Draw your own! Here is one example:

(d) Answers may vary. Draw your own! Here is one example:

(e) Answers may vary. Draw your own! Here is one example:

(f) Answers may vary. Draw your own! Here is one example:

Problem 4.2 (a) acute

(b) right

(c) obtuse

(d) acute

(e) right

(f) obtuse

Problem 4.3 No

Problem 4.5 Yes

Copyright © ARETEEM INSTITUTE. All rights reserved.

Problem 4.8 $75°, 75°$ or $30°, 120°$

Problem 4.9 1

Problem 4.13 (a) $\angle A = \angle B = \angle C = 60°$
(b) $\angle A = \angle B = 70°$
(c) $\angle B = 50°, b = 10$
(d) $\angle C = 70°, c = 10$

Problem 4.15 $75°$

Problem 4.16 5.5

Problem 4.17 $15°$

Problem 4.18 $90°$

Problem 4.20 $115°$

Problem 4.21 48

Problem 4.22 $(-4, 0)$ or $(-4, 2)$

Problem 4.23 $75°$

Problem 4.24 $36°$

Problem 4.25 6

Chapter 5. You Are Special, Right?

Problem 5.1 29

Copyright © ARETEEM INSTITUTE. All rights reserved.

Answer Key to Problems

Problem 5.2 $3\sqrt{2}$

Problem 5.3 6

Problem 5.4 $\sqrt{2}$

Problem 5.5 $1:4$

Problem 5.7   (a) No

  (b) Yes

  (c) Yes

  (d) No

  (e) Yes

Problem 5.8 No

Problem 5.9 34

Problem 5.10   (a) $2\sqrt{2}$ cm

  (b) $3\sqrt{2}$ cm

  (c) 2 in

Problem 5.11 2

Problem 5.12 Yes

Problem 5.13 Yes

Problem 5.14 $\sqrt{13}$

Problem 5.15 $\dfrac{\sqrt{3}-1}{2}$

Copyright © ARETEEM INSTITUTE. All rights reserved.

Problem 5.16 180

Problem 5.17 $12\sqrt{3} - 18$

Problem 5.18 7.6

Problem 5.19 180

Problem 5.20 70

Problem 5.21 36

Problem 5.22 45°

Problem 5.23 420

Problem 5.24 $\frac{1}{2}$

Problem 5.25 15

Chapter 6. Angles Are Special Too

Problem 6.1 Rectangle. Many examples are possible.

Problem 6.2 Rhombus. Many examples are possible.

Problem 6.3 No

Problem 6.4 Interior angle sum: 1260°, Exterior angle sum: 360°, Interior angle 140°

Problem 6.5 Interior angle sum: 3240°, Exterior angle sum: 360°, Interior angle 162°

Copyright © ARETEEM INSTITUTE. All rights reserved.

Answer Key to Problems

Problem 6.6 $\dfrac{1}{3}$

Problem 6.7 No

Problem 6.8 14

Problem 6.9 $82.5°, 37.5°$

Problem 6.10 Yes

Problem 6.11 $119\sqrt{3}$

Problem 6.12 $288 + 288\sqrt{2}$

Problem 6.13 Yes

Problem 6.15 $2 + \sqrt{2}$

Problem 6.16 23

Problem 6.17 $6\sqrt{6}$.

Problem 6.18 $2 + 2\sqrt{3}$

Problem 6.19 12

Problem 6.20 $2 - \sqrt{2}$

Problem 6.21 $\dfrac{1}{3}\sqrt{6} + \sqrt{2} + 2$

Problem 6.22 $1 - \dfrac{\sqrt{3}}{3}$

Problem 6.23 $\frac{1}{4}(3\sqrt{2} - \sqrt{6})$

Problem 6.24 $\frac{3}{8}(2 - \sqrt{3})$

Problem 6.25 $2 : \sqrt{6} : \sqrt{3} + 1$ w.r.t angle ratio $45 - 60 - 75$

Chapter 7. Discovering Areas

Problem 7.1 $4 : 5$

Problem 7.2 $\frac{27}{2}\sqrt{3}$

Problem 7.3 $18(1 + \sqrt{2})$

Problem 7.4 8

Problem 7.5 15

Problem 7.6 (a) 20
(b) Smallest is 18

Problem 7.7 75

Problem 7.8 $1 : 1$

Problem 7.9 $1 : 2$

Problem 7.11 36

Problem 7.12 $\frac{7}{16}$

Answer Key to Problems

Problem 7.13 They have the same area

Problem 7.14 $3:8$

Problem 7.15 9

Problem 7.16 $1:3$

Problem 7.17 27

Problem 7.18 15

Problem 7.19 2 more pairs

Problem 7.20 750

Problem 7.22 $\frac{1}{3}[ABCD]$

Problem 7.23 $6\sqrt{3}$

Problem 7.24 20

Problem 7.25 32

Chapter 8. Conquering Areas

Problem 8.1 $11, 10$

Problem 8.2 12

Problem 8.3 $76, 48$.

Problem 8.4 6

Problem 8.5 $\dfrac{12}{5}$

Problem 8.6 (a) 75, (b) 32, (c) 32, (d) 14

Problem 8.7 24

Problem 8.8 $\dfrac{25}{2}$

Problem 8.9 12

Problem 8.10 $\dfrac{1}{6}$

Problem 8.11 15

Problem 8.12 3

Problem 8.13 (a) 18. (b) 18.

Problem 8.14 $\dfrac{135}{2}$

Problem 8.15 18

Problem 8.16 182

Problem 8.17 5

Problem 8.18 27

Problem 8.19 96

Copyright © ARETEEM INSTITUTE. All rights reserved.

Answer Key to Problems

Problem 8.20 6

Problem 8.21 30

Problem 8.22 (a) $\dfrac{81}{4}\sqrt{3}$

(b) $27\sqrt{3}$

(c) $30\sqrt{3}$

Problem 8.23 $\dfrac{27}{4}$

Problem 8.24 5

Problem 8.25 $\dfrac{1}{2}$

Chapter 9. Circle Around

Problem 9.1 25π; 10π

Problem 9.2 No

Problem 9.3 8π

Problem 9.4 64π

Problem 9.5 (a) Jerry

(b) Both have eaten the same

Problem 9.6 16.

Problem 9.7 (a) 2π, (b) $2\pi - 4$.

Copyright © ARETEEM INSTITUTE. All rights reserved.

Problem 9.8 (a) 24, (b) ≈ 21.

Problem 9.9 (a) $25\pi - 50$, (b) $\dfrac{125\pi}{4} - 50$.

Problem 9.10 (a) 24, (b) ≈ 21.

Problem 9.11 4.

Problem 9.12 10.

Problem 9.13 2.

Problem 9.14 20π.

Problem 9.15 $6 : \pi$

Problem 9.16 (a) $4 + 2\sqrt{2} + 2\pi$. (b) $\pi + 2\sqrt{2}$

Problem 9.17 1.

Problem 9.18 0.

Problem 9.19 10

Problem 9.20 π

Problem 9.21 $6\pi - \dfrac{9}{2}\sqrt{3}$

Problem 9.22 $9\sqrt{3} - \dfrac{9}{2}\pi$

Problem 9.23 $(\dfrac{1}{2}\pi - 1) : 1$

Copyright © ARETEEM INSTITUTE. All rights reserved.

Answer Key to Problems

Problem 9.24 64π

Problem 9.25 $\dfrac{\pi}{6}$

Chapter 10. Circle Back

Problem 10.1 6π in

Problem 10.2 $72°$

Problem 10.3 (a) 2 (b) $\dfrac{\pi}{2}$

Problem 10.4 8m^2

Problem 10.5 $10\pi, 10\pi$

Problem 10.6 $3:4$.

Problem 10.7 350π

Problem 10.8 8

Problem 10.9 $2\pi + 12$.

Problem 10.10 $\dfrac{4\pi}{3}$

Problem 10.11 $\dfrac{1}{2}$

Problem 10.12 50π

Problem 10.13 26 cm^2

Problem 10.14 20π.

Problem 10.15 $\approx 42\%$

Problem 10.16 (a) $4+2\pi$, (b) $4+2\pi$.

Problem 10.17 (a) 8π, (b) 8π.

Problem 10.18 $\dfrac{\pi-2}{2}$

Problem 10.19 $5\sqrt{3}$

Problem 10.20 $3:4$

Problem 10.21 $\dfrac{9}{2}(\pi-\sqrt{3})$

Problem 10.22 6π

Problem 10.23 $\dfrac{25}{2}\pi+12$

Problem 10.24 $\dfrac{3\pi}{2}$

Problem 10.25 4π

Chapter 11. A Whole New Dimension

Problem 11.1 Intersect: $\overline{AD}, \overline{AE}, \overline{BC}, \overline{BF}$ Parallel: $\overline{CD}, \overline{EF}, \overline{GH}$ Skew: $\overline{CG}, \overline{DH}, \overline{EH}, \overline{FG}$

Problem 11.2 (a) 1 ft^3 or 1728 in^3

(b) 7 ft² or 1008 in²

Problem 11.3 (a) 228π

(b) 144π

Problem 11.4 $36\pi, 36\pi$

Problem 11.5 $48; 96$

Problem 11.6 Same volume but different surface areas

Problem 11.7 Same surface area but different volume.

Problem 11.8 13

Problem 11.9 $\sqrt{l^2 + w^2 + h^2}$

Problem 11.10 16π

Problem 11.11 The one ball of radius 4 has more volume

Problem 11.12 The two balls of radius 3 have more combined surface area

Problem 11.13 18

Problem 11.14 $\dfrac{3179}{16\pi}; \dfrac{2057}{8\pi}$

Problem 11.15 5.6π cm²

Problem 11.16 $2\sqrt{5}$

Problem 11.17 $\dfrac{5}{54}$

Problem 11.18 280

Problem 11.19 6000

Problem 11.20 192π

Problem 11.21 $\dfrac{\sqrt{2}}{6}$

Problem 11.22 $20 - 4\sqrt{5}$

Problem 11.23 8

Problem 11.24 120cm^2

Problem 11.25 Yes

Index

Symbols

x-axis 37
y-axis 37
30-60-90 triangle 107
45-45-90 triangle 106

A

acute angle 59
acute triangle 82
adjacent angles 59
alternate exterior angles 60
alternate interior angles 60
altitude 84
angle 16, 58
angle bisector 84
angles in 3D 229

angular size 187
arc 186
arc length 188
area 17

B

ball 231
box 230

C

central angle 187
chord 186
circle 186
circumference 188
complementary angles 59
cone 233
congruent 17
congruent triangles 82

coordinate plane 37
corresponding angles 60
cylinder 231

D

diameter 186
dimension 228
distance formula 105

E

edge 229
equilateral 127
equilateral triangle 17, 82

F

face 229

I

intersecting lines 16
isosceles right triangle 106
isosceles triangle 82

L

length 17
line 16, 58
line segment 17, 58

M

major arc 186
median 84

minor arc 186

O

obtuse angle 59
obtuse triangle 82
origin 37

P

parallel 16, 60
parallel lines in 3D 228
parallel planes 228
parallelogram 17
perimeter 17
perpendicular 58
perpendicular bisector 87
pi 188
plane 228
point 16
polygon 127
pyramid 232
Pythagorean theorem 104

R

radius 186
ray 58
rectangle 17
rectangular prism 230
rectangular pyramid 232
reflex angle 59
regular 127
rhombus 17
right angle 59
right triangle 82

INDEX

S

same-side exterior angles	60
same-side interior angles	60
scalene triangle	82
sector	187
semicircle	186
shape	17
similar triangles	83
skew	228
slant height	236
solid	228
special angle	105, 126
sphere	231
straight angle	59
supplementary angles	59
surface area	229

T

tangent	189
transversal	60
triangle	17
triangle inequality	88

U

unit cube	230

V

vertex	229
vertical angles	59
volume	229

Made in the USA
Columbia, SC
06 July 2024

38004492R00150